Steffen Helln

Killifish

Egg-laying Tooth-carps
Purchase, Care, Feeding, Diseases, Behavior
Special Chapter: Breeding Killifish

With Color Photos by Burkhard Kahl
and Drawings by Fritz W. Köhler

Consulting Editor: Matthew M. Vriends, PhD

BARRON'S

New York • London • Toronto • Sydney

All inquiries should be addressed to:
Barron's Educational Series, Inc.
250 Wireless Boulevard
Hauppauge, NY 11788

Library of Congress Catalog Card No. 90-39368

International Standard Book No. 0-8120-4475-4

Library of Congress Cataloging-in Publication-Data

Hellner, Steffen.
    [Killifische. English]
    Killifish: egg-laying tooth-carps: purchase, care,
feeding, diseases, behavior: special chapter, breeding
killifish / Steffen Hellner; translated from the German by
Rita and Robert Kimber; with color photos by Burkhard
Kahl and drawings by Fritz W. Köhler.]
        p. cm.
    Translation of: Killifische.
    Includes bibliographical references and index.
    ISBN 0-8120-4475-4
    1. Killifish. I. Kahl, Burkhard. II. Köhler, Fritz W. III.
Title.
SF458.K54H4413   1990
639.3'753—dc20   90-39368
                                                    CIP

PRINTED IN HONG KONG

9012   4900   987654321

About the Author
    Steffen Hellner has been keeping aquarium fish for
over 15 years and successfully breeds catfish, cichlids,
and—his specialty—killifish. He frequently gives talks to
aquarists on the care and breeding of killifish and on their
behavior in their native waters.

Cover Photos:
Front cover: *Aphyosemion sjoestedti* (golden pheasant
Gularis) "blue."
Back cover: *Aphyosemion bitaeniatum* "Umudike."
Inside front cover: A pair of *Cynolebias nigripinnis*
(black-finned pearl fish).
Inside back cover: A male *Pterolebias peruensis* (Peru-
vian longfin).

A Note of Warning
    In this book electrical appliances commonly used with
aquariums are described. Please be sure to observe the
safety rules on page 15; otherwise there is a danger of
serious accident.
    Before buying a large tank, check how much weight
the floor of your apartment can support where the tank is
to be located (see page 16).
    Sometimes water damage occurs as a result of broken
glass, overflowing, or a leak in the tank. An insurance
policy that covers such eventualities is therefore highly
recommended.
    Make sure that children (or adults) do not eat aquar-
ium plants. These plants can make people quite sick. Also
make sure that fish medications are always out of reach of
children.

Contents

Preface

Aquarium buffs know right away what is meant when they hear the word "killis." Killifish, or killi, is the name applied to members of the family of egg-laying tooth carps (Cyprinodontidae). For many years killifish were kept pretty much only by fanciers well versed in the art of keeping fish, but by this time many aquarists without very much practical experience aspire to keep these colorful aquarium dwellers, too. Unfortunately, their first attempts often end in disappointment because neophyte aquarists tend not to realize that success is possible only if the keeper knows the different species and—equally important—their particular requirements. This Pet Owners' Manual will help you avoid mistakes in maintenance and care from the very outset. Steffen Hellner, who has been keeping and breeding different killifishes for the past 16 years, answers all the major questions related to the care of these fish. His advice is based not only on his own experience but also on information gathered from other killifish experts.

The author explains in detail why it is important to decide before buying fish whether you want to set up a species tank or a community tank; what to watch out for when purchasing fish; how an aquarium that is to house killifish should be set up; and how the fish are acclimated properly. Of course, he also provides all the necessary information on water quality, filtering, heating, and lighting. His advice and precise instructions are easy to follow even for beginners, and there are informative drawings that illustrate basic procedures of aquarium care as well as fascinating aspects of killifish behavior.

The extensive descriptions in the section on different killifish species—containing data on distribution, habitat, and life pattern, as well as specific instructions on maintenance and breeding—will help you choose the kinds of killis that are right for you and to combine compatible species. The author's advice on creating exemplary killifish communities and combinations of killis with other types of fish is also very helpful.

Proper nutrition is extremely important for the well-being of killis. Since these fish depend primarily on live food in nature, they obviously have to be given a similar diet in an aquarium. The author explains exactly how killifish in an aquarium can be provided with a proper diet.

If killifish are kept and fed properly they very rarely get sick, but if a disease should turn up in your aquarium, you will find information and advice on what to do in the chapter Diseases of Killifish.

Anyone who has killifish usually likes them to produce offspring sooner or later, because for many aquarium enthusiasts the crowning achievement of their hobby is to raise baby fish. This is why a special chapter is devoted to raising killifish—a field that is a specialty of the author's and one in which he has scored impressive successes. This chapter includes comprehensive instructions tested by the author and easily implemented by anyone.

Excellent photographs, taken especially for this Pet Owner's Manual by the aquarium specialist Burkhard Kahl, depict killifish in all their colorful splendor.

The author of the book and the editors of Barron's Series of Nature Books wish you much enjoyment of your killifish.

The author and the publisher wish to thank all those who had a share in producing this book: Godja Worms, Rainer Brebeck, Eckhard Busch, Norbert Dadaniak, Wolfang Eberl, Wolfgang Grell, and Bernd and Klaus Schölzel for their helpful contributions; the photographer Burkhard Kahl for his beautiful color photographs; Fritz W. Köhler for his informative drawings; and Harald Jes, director of the Aquarium at the Cologne Zoo, for checking the chapter Diseases of Killifish.

Useful Facts About Killifish

Introduction to the Family

Aquarists refer to the family of egg-laying tooth carps, or Cyprinodontidae, as killifish, or "killis," for short. It would seem at first glance that this name was derived from the verb "to kill," especially since all species of killifish used to have the reputation—unfounded as it turns out—of being quite aggressive toward others of their kind. However, the name has a quite different origin. In the sixteenth century many Dutch immigrants settled along the East Coast of America. These settlers used the Dutch word "kil" for small bodies of water and quite naturally called the small fish they found in these waters "kil-vissen," or "fish of the kils." Among these small fish there were egg-laying tooth carps of the species *Fundulus heteroclitus*. These minnows are called "common killifish" even today. The name killifish acquired widespread use when the American Killifish Association (AKA) was founded in the early 1960's. Many associations in other countries have also adopted this term, among them the British Killifish Association (BKA), established in 1968, and the German Killifish Association, which was established in 1969.

Where Killifish Are Found

Killifish are found all over the world, with the exception of Australia and the Arctic regions. The greatest variety of species developed in the tropical waters of Africa and South America. But killifish also live in subtropical and temperate latitudes because they have developed some unique strategies of survival. In the Mediterranean area, fish of the *Aphanius* genus (see page 59) are found along the coast and in oases of the Sahara as well as in the salty lakes of Turkey. Killifish even populate waters that are several times saltier than seawater. Perhaps most famous for their adaptability to extreme living conditions are the North American desert fish of the genus *Cyprinodon*. Some species of this genus live in water temperatures that are too hot for any other organisms except some algae. In Mexico, *C. pachycephalus* lives in waters of 99° F (37° C) and *C. nevadensis* holds the record of surviving 110° F (43° C)!

How Killifish Found Their Way into Aquariums

Toward the end of the nineteenth century aquarists were familiar mostly with North American killifishes and with a few African species. Sailors caught the fish for interested fanciers and brought them back from their distant travels under most adverse conditions. In the twentieth century the hobby of keeping fish gained greatly in popularity, and there was a demand for more and more different kinds of killifish. Dealers and scientists got together to organize regular imports, and with the founding of specialized fanciers' associations (see Addresses, page 69) the popularity of keeping killifish reached new heights.

Why the Scientific Names Are Important

When killifish fanciers refer to fish by their scientific (Latin and Greek) names this is not an affectation or elitist. The scientific names are so unambiguous that they make accurate communication between fanciers from different countries possible. Vernacular names are not officially defined anywhere; they are often changed and easily misunderstood. Only a few have acquired general currency, such as lyretail, lyretailed *Panchax*, or Cape Lopez lyretail for *Aphyosemion australe*. In 1766, Carl von Linné (Carolus Linnaeus) intro-

Useful Facts About Killifish

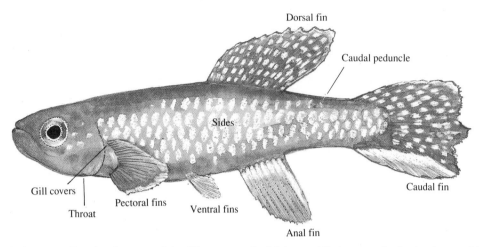

Dorsal fin

Caudal peduncle

Sides

Caudal fin

Gill covers

Pectoral fins

Throat

Ventral fins

Anal fin

Fish anatomy. Knowing the names of the different parts of a fish is especially important for fanciers interested in the behavior and life patterns of killis.

duced the so-called binomial nomenclature, and since then every organism—whether plant or animal—has been classified according to this system.

The first word in the name is capitalized and denotes the genus of the organism; the second word, which is lowercase, indicates the species. For example; in *Aphyosemion gardneri*, for the steel blue *Aphyosemion*, *Aphyosemion* is the genus and *gardneri* is the species.

Species can be further divided into subspecies. *Aphyosemion gardneri*, for instance, exists in several forms. The nominate subspecies, in this case *A. gardneri gardneri*, is the form that was first described and therefore repeats the species name. Another subspecies is called *A. gardneri nigerianum*. These distinctions are of special importance to fanciers who want to breed their fish (see Breeding Killifish, page 44). Sometimes fish of the same subspecies are caught in several different locations. The difference between these fish from different geographic areas—differences in color, for instance—can be striking or barely apparent. To

accurately indicate such differences, the name of the location of origin is included in the scientific name. Thus, *A. gardneri nigerianum* Makurdi gives the genus, species, subspecies, and the place of origin—Makurdi, in Nigeria. In scientific literature, the name of the person who first described the organism and the year of the first description are added, too.

Note: After the first mention, the genus is often abbreviated to the first letter, especially in lists or when a text focuses exclusively on one genus (see examples in preceding paragraph).

The Life Patterns of Killifish

The life patterns of killifish vary quite a lot, and keepers of these fish should be familiar with the most important differences.

Annual fish, such as the genera *Cynolebias* and *Pterolebias*, live in waters that dry up at certain times of year.

The life cycle of these fish is compressed into less than a year, and they rarely live longer than nine months. Annual fish lay their eggs on the bottom, and because of this they are called bottom spawners. When the water evaporates, the bottom, where the spawn lies, remains moist relatively long, and the eggs have a chance to develop by the time the rainy season returns.

Later, when the bodies of water fill again, the young fish hatch. By the time they are four to six weeks old they are ready to reproduce, and at three to five months they have grown to their full size. They die when the waters dry up again, but the survival of the species is assured.

Semiannual species, like *Paraphyosemion* and *Aphyosemion walkeri*, inhabit waters that do not dry up regularly but may carry water year-round.

These species deposit their spawn either on the bottom or in the bottom mud. These killifish, too, are called bottom spawners. Their eggs develop either in the water or, during dry periods, in the damp ground. Depending on how often the water dries up, these semiannual fish can live as long as several years.

Nonannual species, such as *Epiplatys* and small species of the genera *Aphyosemion* and *Roloffia*, live in bodies of water that carry water all year, and they deposit their spawn on aquatic plants to which the eggs adhere.

Because of this reproductive behavior, killifish of this type are called plant spawners. The eggs develop in the water but can withstand short dry periods. This type of fish, which by the way includes the majority of all killifishes, lives up to five years and disproves the notion that all killis are short-lived.

If conditions in an aquarium are optimal for a given species of killifish (see page 15), the fish can easily live to their full life span. In nature, by contrast, conditions are often unfavorable, so that killis die prematurely. They also often fall victim to predatory fish.

Social Behavior

Apart from the method of reproduction, very little is really known about the behavior of killifish in nature. The majority, except for the lamp-eyes (see descriptions, page 60), are solitary; lamp-eyes form small shoals.

Fights to Defend Territory: Several species of killifish are often kept together in one tank. The brightness of coloration provides a rough indication of the relative compatibility of different species. Fish that live in shoals usually have quite inconspicuous coloring; species that are decidedly unsociable, on the other hand, such as the *Nothobranchius* species, are very brightly colored. If several specimens of this latter type are placed in an aquarium all at once, territorial fights immediately erupt, especially between males (see drawing, page 8). The fish swim toward and circle each other with fully spread fins and wide-open gill covers. This intimidating behavior is often enough to rout one of the rivals. Minor injuries are sometimes incurred, but rarely anything more serious than a torn fin. During transport home in small plastic bags or if the aquarium is too small, it can happen that two males attack each other so ferociously that the weaker is killed (see Transporting Killifish, page 26).

Courtship Behavior: All killifish act very similarly during courtship. If a male has discovered a female, he swims around her with his fins fanned out wide (see drawing, page 52). This fanning of the fins alternates with snakelike movements of the body during which the fins lie close to the body. If you watch some of the small *Rivulus* species in particular, you see the male come to a sudden stop and move his whole body, especially the head, in a jerky fashion. If the female is not ready to spawn, she responds with flight; otherwise she moves closer to the male.

Reproductive behavior: Males of species that spawn on plants or on the bottom try, if their courting has been accepted, to swim directly above or next to the female, which they then "grasp" with their dorsal and anal fins and try to press against the place where the eggs are to be laid, the spawning substrate (see Spawning Aids, page 47). As they spawn, both fish tremble violently and push away from each other and from the spawning substrate. These movements cause the eggs to be hurled against or into the substrate.

The reproductive behavior of annual fish that dive into the bottom mud to spawn differs from this pattern. After successful courtship, both partners move close to each other. Observations on my part over several generations of many *Pterolebias* and *Cynolebias* species indicate that it is the female that determines the spot chosen for spawning. The males of these species repeatedly point their heads in the direction of the bottom or touch it with the head, but diving occurs only if the female "accepts" the indicated spot or selects a different one. Whereas spawning takes only seconds with species that spawn on plants or on the bottom, pairs that dive into the bottom stay buried up to 15 minutes (sometimes even longer) to spawn. Usually the two partners re-emerge from the ground separately. Particularly if you keep *Pterolebias*, you sometimes see other females that are ready to spawn follow a pair that has disappeared into the bottom. I have frequently seen such females actually spawn. They thus manage to get their eggs fertilized without first engaging in a courtship ritual with the male. Whether this also occurs in nature is not known.

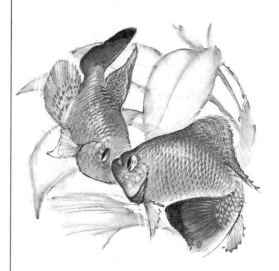

Intimidating behavior. Brightly colored males, such as members of the *Nothobranchius* genus depicted here, commonly engage in fierce fights over rank order. However, the intimidating posture of raised fins and the gill covers is often enough to drive a contestant away.

Depositing eggs in the bottom can be regarded as a kind of "incubation protection." Real brood care, however, as some cichlids and catfish practice it, has thus far not been reported in killifish.

Advice for Buying

Species Tank or Community Aquarium?

In a species tank only fish of one species, for instance, *Aphyosemion gardneri*, are kept. In contrast, several different, compatible species live together in a community tank. The community might consist of several killifish species or of a combination of killis and some other suitable species. To decide which type of tank to set up, you must study and compare the characteristics and needs of individual killifish species given in the descriptions starting on page 53. Make sure to pay special attention to the following:
• the size of the fish
• their behavior toward members of their own as well as of other species
• the conditions of their natural habitat
• the geographic origin of individual species

Size of the Fish: Killifish that differ in size should not be put together in the same aquarium. The highly aggressive but small *Aphyosemion (Diapteron) fulgens,* for instance, cannot hold its own against the considerably larger *A. gardneri.* If you introduce other kinds of fish (see page 14) into a tank with killis, these, too, must match in size. Small differences do not matter.

Behavior: The behavior patterns of killifish vary depending on the species (see page 7). In deciding between a species and a community aquarium you must also take into account the assertiveness of the species you are considering. Males of the species *Cynolebias nigripinnis* and of many other South American bottom spawners are highly aggressive toward each other, but fights hardly ever result in serious injuries. However, if they are combined with fish of the same size belonging to the *Aphyosemion gardneri* group or to the genus *Rivulus*, there may be fatalities because the flight instinct is not well developed in *Cynolebias nigripinnis*. In nature these killis disappear into the bottom when they sense danger, but they cannot do this in a community tank. And since *Cynolebias nigripinnis* males constantly try to intimidate other fish, they are often literally torn to pieces. This problem does not arise if they are combined with surface-dwelling fish, such as *Epiplatys* and *Aplocheilus*, or with lamp-eyes. The larger *Nothobranchius* species usually hold their own in a community with medium-sized *Aphyosemion* species (for further information on the subject of compatability see Exemplary Communities, page 13).

Conditions in the Natural Habitat: These are of major importance when you set up a community tank. Only if the environmental requirements are the same should different species be combined. For instance, both *Aphyosemion splendopleure* and *A. ogoense* are native to Gabon, but they are not compatible. *A. splendopleure* lives in waters in the coastal plains that have a temperature of around 75° F (24° C). *A. ogoense* comes from the Bateke Plateau, where the water is considerably cooler, between 66 and 70° F (19°–21 C). *A. filamentosum* from Togo and Nigeria is a much better bet for a tank containing *A. splendopleure,* and *A. bualanum* from the highlands of Cameroon is a better match for *A. ogoense* (see descriptions of some of these varieties, page 53).

Geographic Origin: The mere fact that different fish come from the same continent or from overlapping ranges is no great aid in deciding whether they should be kept in separate tanks or in a community tank.

In nature, individual species live spread out over large areas and can avoid each other. Conditions in an aquarium are very different. The fish must coexist in a confined space and have no chance to escape from each other.

An aquarist who wants to create an aquarium with certain typical geographic characteristics would, of course, like to include killifishes from the same or similar geographic areas. If you are in this situation, make doubly sure that the species you pick are well matched in terms of the selection criteria already discussed.

Advice for Buying

Where You Can Get Killifish

It is sometimes not easy to find a certain kind of killifish, but if you explore all available sources you will probably end up with the killi species you wanted for your aquarium.

Pet Dealers: Many well-run pet stores have killifish for sale, but unfortunately the selection still tends to be rather limited. Only *Aphyosemion gardneri, A. australe,* a captively bred, golden version of the same species (Cape Lopez "gold" lyretail), *A. sjoestedti,* and some *Nothobranchius* and *Epiplatys* species are regularly available.

American Killifish Association: Through this organization you can obtain an almost unlimited selection of species. Contact the AKA (see Addresses, page 69), and visit one of the many regional shows. Dates and places are listed in various aquarium magazines (see Addresses, page 69). If you join the AKA, you will meet experienced killifish keepers and breeders who will be able to assist you with theoretical and practical advice.

Breeders: Some breeders belonging to the AKA ship fish they have raised by mail or other carrier. You can get their addresses through the AKA.

Specialized Magazines: In these publications (for instance the AKA publication) you will often find rare killifish advertised for sale. The ads are usually placed by pet dealers specializing in the aquarium trade. The dealers may be located at some distance from you, but they generally mail fish. Although buying killifish sight unseen is not recommended as a rule, my experience has on the whole been positive.

The Best Time to Buy

The time of year and your vacation plans should be taken into consideration when you buy killifish.

Time of year: If you order fish through the mail or want to ship fish yourself, you must keep in mind outdoor temperatures. Fish rarely survive temperatures colder than 50° F (10° C) or warmer than 77° F (25° C) without damage. If you transport the fish yourself, the time of year does not matter so much as long as you take proper precautions (see Transporting Killifish, page 26).

Vacation time: If you can, you should buy your fish at least two months before a planned vacation so that the fish have time to acclimate. Be sure to arrange for a knowledgeable person to look after the fish if you have to go away. Fully grown fish can go without food and care *for a short time,* but young fish must be fed daily and cannot be left unattended. Also keep in mind that fish sometimes get sick even if they receive good care (see Diseases, page 40). In such a case, the caretaker must be experienced enough to be able to initiate treatment to prevent major losses.

Tip: My purchases of killifish have been most successful when I bought them right after summer vacation. At this point—also before holidays like Easter and Christmas—the selection at pet stores is greatest.

What to Watch Out for

Buying a living creature is always a matter of trust and something of a gamble. Disappointment can be kept to a minimum if you yourself are able to judge the condition of the killis you consider buying.

Is This Killifish Healthy?

It is important to assess the state of health of a fish correctly. To do this, look at the fish from the side as well as, always, from above (see drawing on page 11). Take an especially good look at the body, skin, fins, eyes, gills, and throat.

Advice for Buying

Body: In a healthy killifish the outline of the head and body—seen from above—forms a very slight arch that ends in the caudal peduncle (the base of the tail). If the head seems noticeably set off from the body and the body is considerably narrower than the head, or if the caudal peduncle looks swollen, you should not buy the fish.

Healthy, well-fed killifish have a smooth, somewhat convex abdominal line. In no case should the height of the body be less than that of the head; such fish inevitably die. Do not buy fish with bloated bellies, either, or fish that look emaciated.

Skin: The skin should be free of films or coatings of any sort and should not look cloudy. Diseases caused by parasites (see Diseases, page 40), some of which manifest themselves in small white dots or in white, fuzzy growths (mycoses), are usually visible at a glance. The scales should cover the body smoothly, not stick up.

Fins: All the fins should be present and well formed. Do not buy killifish that swim around with their fins clamped to the body or that have partially missing or malformed fins. If the fins are frayed even though the aquarium contains no fin-nipping fishes (such as Sumatra barbs or jewel tetras), this may be a first sign of tuberculosis (see page 42); also check to make sure the fish do not have bulging eyes!

Eyes: Both eyes should be of the same, normal size. The selectively bred golden strain of *Aphyosemion australe* (Cape Lopez "gold" lyretail) tends to have somewhat atrophied eyes. Often one eye is obviously smaller than the other, or both eyes may be abnormally small. If you look at all the Cape Lopez lyretails swimming around in a tank, you will quickly recognize the abnormal ones. Also reject fish whose eyes protrude!

Gills: A healthy fish moves its gill covers in a definite rhythm. If the gill covers stay permanently lifted, this suggests sickness.

Throat: If the throat is swollen between the gill covers, this suggests thyroid problems (see page 42).

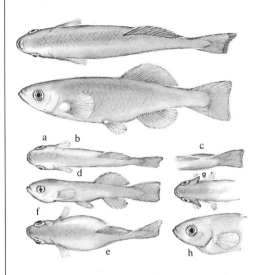

State of health. Before buying a killifish, look at it from above and from the side. If (a) the head, viewed from above, is set off markedly, (b) the body is considerably narrower than the head, and (c) the caudal peduncle is swollen, you should not buy the fish. Clamped fins (d), bulging eyes (f), gill covers that stay open (g), a bloated abdomen (e), and swelling of the throat (h) also suggest illness.

The Age at Purchase

Even though it is not always easy to assess the age of killifish, there are a few signs that may help.

It is best to buy fish that have not quite reached their full size.

The size of a fish will tell you whether it is a juvenile or an "old" specimen—assuming, of course, that you know how large the species in question gets (see descriptions, page 53).

If the head is visibly set off from the body at the neck, you are looking at a fully grown fish (see drawing on right). This sign is helpful because killifish are sometimes stunted in their growth.

Old fish are often very thin, and their skin is not

as smooth as that of juvenile fish.

In annual fish, the intensity of color diminishes dramatically with age and the body takes on a glassy look.

Distinguishing the Sexes

With killifish it is easy, quite early, to tell males from females.

Females are either a uniform gray to brown, or their coloration is much more muted than that of the males.

Males are brightly colored, which clearly sets them off from the females.

Important: If fish belonging to small *Rivulus* species (see descriptions, page 53) are taken out of the aquarium, they often lose their color in a reaction of fright. It is then very difficult to tell what sex they are unless you are very practiced. However, this loss of color is only temporary. Once the fish calm down, the normal colors return.

Should You Buy One Fish, a Pair, or a Shoal?

This decision depends not just on the particular life pattern of the species you are considering (see descriptions, page 53) but also on your plans for the fish.

Single specimens: If you are not planning to breed the fish, you can introduce single individuals of a species into a community tank. In that case you will probably opt for a male because males are more colorful than females. Females, too, can live singly in an aquarium, however.

Pairs: It is of course much more interesting to keep at least one pair of a species in a community aquarium because the courtship and spawning

behavior of killis is interesting to watch (see Breeding, page 44). Anyone hoping to breed fish should buy at least a trio (one male and two females) of the same species.

Determining a fish's age. Look at the head and back of a killifish from the side. The outline of a juvenile fish is smooth (above), whereas in mature fish (below) there is an indentation in the line joining the head and back.

Important Note: Make sure when you buy fish that the males and females are more or less the same size. Especially in killis belonging to the *Aphyosemion cameronense* group the females sometimes are larger than the males (especially *A. mimbon*). Females that are larger than the males are known to display extremely aggressive behavior toward them!

A Shoal: Lamp-eyes (see descriptions, page 60) are shoal fish or at least fish that live in social groups. If you want to keep these species, you should buy at least six to ten fish, and even if you do not plan to breed them, such a group must include females in a proportion of about one female for every two males.

Advice for Buying

Exemplary Communities

Group	Tank size	Water	Food	Setup, plants	Fish
Small killifish from the West African rainforest	24 inches (60 cm) or more	4–10° dH 6° dCH, pH 6–7, 66–73°F (19–23° C)	Live and frozen food	Dark, soft bottom material, root wood; *Bolbitis, Anubias, Ceratopteris* (floating)	Eight lamp-eyes (*Procatopus, Plataplochilus*) and a total of five pairs of *Aphyosemion cameronense, A. ocellatum,* or *A. ogonese.*
East African annual fish	32 inches (80 cm) or more	4–25° dH, pH up to 7.5, 75–79°F (24–26° C)	A few food flakes, frozen food, live food (*Tubifex!*)	Eel grass around edges, floating plants, soft bottom material, root wood	20–25 males of various *Nothobranchius* species; keep females separate!
South American bottom spawners	40 inches (100 cm)	4–15° dH, pH 6–7.5, 68–79°F (20–26° C)	All kinds of food	Root wood; sandy bottom; *Echinodorus,* water pennywort, eel grass, floating plants	Two to three pairs each of various killis, adding up to at most 20 fish; add two males and three to five females of *Apistogramma borelli* or *A. agassizii* and six armored catfish
Large killifish from West African savannas	40 inches (100 cm), preferably larger	4–20° dH, pH 6–7.5, 72–79° F (22–26° C)	All kinds of food but not too much (obesity); lots of water fleas	Thick plantings along edges, Egyptian lotus, eel grass, *Najas, Ceratopteris*	No more than a total of nine pairs of the species *Aphyosemion sjostedti, A. fallax, Roloffia toddi, R. occidentalis*; add two pairs of *Pelvicachromis taeniatus* or *Nanochromis nudiceps*
Small *Rivulus* species	32 inches (80 cm)	4–15° dH/ 8° dCH, pH 6–7, 72–77° F (22–25° C)	Live and frozen food, little dry food	Thick plantings along edges with water pennywort, *Echinodorus nodorus latifolius* and, on top, hornwort or *Ceratop*teris, root wood	Five to six pairs each of *Rivulus agilae* and *R. strigatus*; six to eight armoredl catfish and bottom-dwelling characins (*Characidium* genus); six pencilfish (*Nannostomus*)
Southeast Asian killifish	40 inches (100 cm)	4–15° dH/ 8° dCH (because of the companion fish), pH 6–7.5, 75–86° F (24–30° C)	All kinds of food	Thick planting, water trumpets, *Ceratopteris*	Two to three pairs of *Aplocheilus lineatus and A. panchax*; add two or three pairs of *Betta imbellis* or *B. splendens,* five loaches, and ten nonaggressive barbs (*Rasbora*)

Advice for Buying

Who Gets Along with Whom?

Killifish can be combined with other kinds of fish that are compatible with them.

American Characins

Genera/Species: *Hyphessobrycon, Hemigrammus, Nannobrycon,* and *Nannostomus,* as well as neon tetras (*Paracheirodon innesi*), cardinal tetras (*Cheirodon axelrodi*), and false rummynose tetras (*Petitella georgiae*).

Life Pattern: Shoal fish (keep at least six fish) for tanks holding at least 15 gallons (60 L).

Food: Live, frozen, and dried.

Water: 4–15° dH; pH 6–7; 72–82° F (22–28° C).

Companion fish: Small to medium-sized killifish, such as *Rivulus, Aphyosemion, Epiplatys,* and *Aplocheilus.*

Armored catfishes

Species: *Corydoras paleatus, C. erhardti,* and *C. barbatus.*

Life pattern: Shoal fish, for tanks holding at least 20 gallons (80 L).

Food: Omnivorous; particularly like water fleas, worms, and food tablets.

Water: 4–10° dH; pH 6.5–7; 65–75° F (18–24° C).

Companion fish: Lamp-eyes and small Rivulus species.

Aspidoras Species and Corydoras Species from Warm Areas

Species: *Aspidoras pauciradiatus, Corydoras aeneus, C. sterbai, C. rabauti,* and *C. julii.*

Life pattern: Shoal fish. Species up to 1.5 inches (4 cm) for tanks holding at least 10 gallons (40 L); aquariums for larger species must hold at least 20 gallons (80 L).

Food: Omnivorous.

Water: 4–15° dH; pH 6–7.5°; 72–82° F (22-28° C).

Companion fish: All killifish measuring up to 4 inches (10 cm), especially all *Paraphyosemion* and large *Rivulus* species. Can also be combined with *Nothobranchius.*

Various Catfish

Genera/Species: Antenna catfish (*Ancistrus*) and mailed catfish (*Loricaria* and *Rineloricaria*), as well as *Sturisoma, Farlowella,* and *Otocinclus.*

Life pattern: Solitary; need cavities to hide in. For tanks holding at least 20 gallons (80 L).

Food: Dry food and food tablets made of plant matter: spinach, lettuce, and dandelion greens.

Water: 4-15° dH; for *Ancistrus,* up to 25° dH; pH 6–7.5°; 68–79° F (20–26° C), for *Ancistrus* up to 86° F (30° C)!

Companion fish: Medium to large killis; *Otocinclus* can also be combined with smaller killis, such as *Diapteron* and *Pseudepiplatys.*

Dwarf Cichlids

Genera: *Apistogramma, Papiliochromis,* and *Pelvicachromis.*

Life pattern: Solitary; territorial; aggressive at spawning time, otherwise peaceful. For tanks holding at least 20 gallons (80 L).

Food: Frozen and live; small amounts of dry food.

Water: 4–15° dH; pH 6–7°; 68–79° F (20–26° C). *P. ramirezi* requires 75–82° F (20–28° C).

Companion fish: All medium to large killifish. Keep no more than one pair in tanks of up to 25 gallons (100 L); larger tanks can include several pairs.

The Right Aquarium for Killifish

If you choose the right kind of aquarium and set it up properly, you are providing your killifish with an environment in which they will feel happy and give you pleasure for a long time. In addition to the tank itself you will need a few technical devices.

Safety Precautions

To create a suitable environment for fish and plants you must make use of several electrical appliances, namely a heater, a filtering system, and lights, and it is important to be aware of the dangers inherent in handling electrical equipment and wires, particularly in combination with water. Be sure, therefore, to observe the following safety rules:

• The electrical appliances described in this book should be UL approved.

• Equipment that will be run *inside* the aquarium must carry a label saying that it is designed for use in water.

• Disconnect all wires before you start work inside the aquarium, or take the equipment out of the aquarium.

• We also urge you to purchase a circuit breaker, equipped with four outlets, that will cut off the electricity if a device or a wire is damaged.

Which Aquarium Is the Right One?

Material: I recommend that you buy a glass tank. There are two types: aquariums with non-rusting frames made of anodized aluminum, plastic, or stainless steel, and frameless all-glass tanks. Both types use silicon glue as sealer. I use plastic tanks only for breeding fish (see Breeding, page 44) because they are easily scratched.

Size: The dimensions of an aquarium depend on the size, number, and requirements of its occu-

pants. The following is a general guideline. Figure on 1.5 to 2 quarts (1.5–2 L) of water per $^3/_8$ inch (1 cm) of fish length (based on adult size). The figure you come up with indicates the actual amount of water but does not include space taken up by bottom material and decorations. You can measure the water by using a calibrated bucket or a watering can when filling the tank. This rule of thumb is very helpful for planning the size of tank to buy, and you can apply it later to check if the aquarium is overpopulated.

Important Note: If you are planning to have a large aquarium setup, ask the architect of your building (sometimes the landlord can tell you) how much weight the floors of your apartment are designed to support. One liter of water weighs 1 kilogram. Therefore an aquarium measuring $40 \times 20 \times 36$ inches ($100 \times 50 \times 40$ cm) weighs 440 pounds (200 kg), since it holds 53 gallons (200 liters). Especially people living on upper floors or in old buildings must be careful about checking the strength of their floors.

Aquarium lights: Mercury vapor lights are suspended above the tank and therefore do not interfere when you work in the tank.

The Right Aquarium for Killifish

Format: Tanks for killifish should, if possible, be wider than high because killifish live primarily in shallow waters in the wild. This applies not only to species tanks but also to community aquariums.

Depending on the size and number of the fish, a species tank should be at least 20 × 12 × 10 inches (50 × 30 × 25 cm) or 24 × 16 × 12 inches (60 × 40 × 30 cm); the minimum measurements for a community tank (see table, page 13) are 32 × 16 × 16 inches (80 × 40 × 40 cm) or 40 × 16 × 16 inches (100 × 40 × 40 cm).

Cover: Killifish can jump well and will escape through the tiniest cracks. This is why a tightly fitting cover is crucial. Even the smallest openings, as around the filter entrance, must be blocked with foam rubber. Most aquariums are sold with a hood with a built-in light fixture, but simple glass covers also work well.

The Right Location

To prevent massive formation of algae, an aquarium with killifish should always be placed away from windows, preferably against the opposite wall. Thanks to modern lighting technology, this no longer presents a problem. Just be sure that there are one or more electric outlets nearby.

Heating

Theoretically, heating the aquarium is not absolutely necessary for most killifish species since rooms that are lived in are normally kept around at least 65° F (18° C). This water temperature is adequate for many killis. A heater is still a good idea because it keeps the water temperature constant. A few killifish species require relatively high temperatures, and for these a heater is crucial. Exact temperature requirements are given under the different species descriptions starting on page 53.

There are several different ways to heat an aquarium.

Thermostatically controlled heaters are made by many manufacturers and sold at pet and aquarium stores. They are installed inside the tank. Choose one in a natural color (such as green or brown) so that it is not too much of an eyesore.

Tip: It is important with any heater to check several times after turning it on that it actually produces the desired water temperature, for the thermostats are not always very accurate. To prevent overheating, the heater should be placed near the filter outlet (water current).

Heating elements are installed in the filtration system, but this presupposes that you have an outside filter (see page 19). Heaters installed in the filtration system regulate water temperature very evenly; they are easy to run and ideal from an esthetic point of view (no cables inside the tank). There are also external filters with heating elements already built-in (available at pet and aquarium stores).

Heat cables, preferably run on low-voltage current, are installed on the bottom of the tank (beneath the bottom material). Unlike water heaters in the tank or in the filter, heat cables warm not only the water but the bottom material or substrate as well. This helps plants thrive, since all aquarium plants like to have "warm feet." In addition the bottom material is slightly aerated, which keeps the plants from rotting.

For large community tanks holding 40 gallons (160 L) or more, this method of heating is ideal. The only disadvantage is that the tank must be emptied if there is a malfunctioning. Fortunately heat cables rarely develop problems.

Killifish of the genus Roloffia. Above: A pair of *Roloffia geryi* Abuko. Below: A male *Roloffia toddi* Barmoi.

The Right Aquarium for Killifish

Filtration

Filtration is crucial for maintaining stable conditions in an aquarium. The filter collects uneaten food, fish excreta, and rotting plant parts. Still, even the best filtering system is no substitute for regular replacement of the water (see page 29). Filters help keep the water quality adequate for the fish between water changes, however.

Internal filters work well for small tanks holding anywhere from 2.5 to 25 gallons (10–100 L). Most internal filters are operated with diaphragm pumps (air pumps). I especially recommend internal foam filters. They are easy to clean, inexpensive, and effective. They are particularly good for rearing tanks (see Breeding, page 44) because the foam cartridges prevent tiny fry from being sucked in. Plastic filters are equally practical but somewhat harder to clean. They have one advantage, however: all kinds of filtering materials can be used in them (peat, crushed lava, tiny clay tubes, porous ceramic filters, filter wadding, and activated charcoal). In rearing tanks, half-filling these filters with peat has proven useful (see pH value, page 23). Internal plastic filters, too, keep the fry from being sucked up.

Tip: I use crushed lava (available at pet and garden supply stores) as filtering material in interior plastic filters, topping it with about $^3/_4$ inch (2 cm) of filter wadding. Such a filter keeps the water crystal clear.

External filters are used primarily in larger tanks (over 25 gallons or 100 L). I have successfully used external power filters (for example, Dynaflo, by Rolf C. Hagan). They are set up outside the tank (behind it, for instance, or in a cabinet underneath). Because external filters do not take up space inside the aquarium, they can be larger than internal filters. External filters are cleaned outside the tank and can be used with different kinds of filtering material. For practical reasons these filters are equipped with hose couplings that allow the filter to be disconnected easily from the circulating system.

Tip: I use the following materials in my external power filter. I place 2 inches (5 cm) of coarse filter wadding over the bottom sieve; this is topped with enough of a somewhat finer filtering material (such as tiny clay pipes or porous ceramic) to leave room for about $^3/_4$ inch (2 cm) of fine filter wadding. If you want the water to be filtered through peat or charcoal, leave enough room for it to fit underneath the fine wadding. Filtering peat, thoroughly rinsed after use in a filter, makes a good spawning substrate for bottom-spawning fishes (see Breeding, page 44). To protect the peat from becoming too clogged, I put an extra layer of fine filter wadding underneath it inside the filter.

Lighting

Most killifish species do best with only moderate illumination. Desert fish and a few species from savanna areas (see descriptions, page 53) like stronger light and sometimes a few hours of direct sunlight.

The length of artificial lighting should, as a rule, be about 12–14 hours a day and be regulated with an automatic timer. You can use mercury vapor lights, although these are not common in the United States. If you can find them, they can be placed on the aquarium or suspended about it. Check with your local retailer to learn which products are available.

Fluorescent light tubes are what most aquarium covers are designed for. If there is room for only one

Killifish of the genus Rivulus. Above left: A male *Rivulus amphoreus*. Above right: A pair of *Rivulus spec. aff. holmiae*. Middle left: Two male *Rivulus agilae*. Middle right: A hermaphroditic *Rivulus ocellatus ocellatus* Grota Funda caught in the wild. Below left: A male *Rivulus luelingi* "Joinville." Below right: A pair of *Rivulus caudomarginatus* Grota Funda caught in the wild.

tube, I use the colors 11 (white) or 32 (warm tone). Types 36 and 77 are also good light sources. If there is room for several tubes, combinations of these colors have shown good results. However, fluorescent light tubes deteriorate quickly, and the light intensity diminishes about 50% within one year. The human eye and the fish hardly notice this decline in brightness, and it is of no great consequence to the well-being of killifish. If you want your plants to grow well, however, the tubes should be replaced because many plants are very sensitive to sudden changes in light intensity or color. If there are two tubes, one should be replaced alternately with the other every six months; if there are three replace one every four months. This keeps the light intensity fairly constant.

Important: I strongly urge you to stay away from "plant lights" with their heavy reliance on the red sector of the color spectrum. Quite apart from unnatural colors these lights produce, they have led to excessive algae growth and stunted plants in my tanks, even when used in combination with other light colors.

High-pressure quartz lamps (HQL) emit a very natural light and come on gradually after being switched on (thus not scaring the fish!). In my opinion there is no better light source for tanks at least 40 inches (100 cm) long. This kind of light is beneficial to the plants in the aquarium as well. Quartz lamps are somewhat more expensive than fluorescent lights, but the bulbs can be used until they wear out without significant loss of light intensity. I have been most pleased with bulbs in warm tones. White bulbs give off a glaring light and encourage the growth of algae.

Burnt-out bulbs should be disposed of as hazardous waste.

Tip: I light my community tank (80 gallons or 300 L) with two 125-watt high-pressure quartz lamps, which the fish (*Rivulus* spec. aff. *holmiae, Aphyosemion sjoestedti, Roloffia toddi*) tolerate without acting shy, which brings out the full palette of their colors. If you keep small *Aphyosemion*

species and other shy killifish, the water surface should be partially covered with floating plants (see Aquarium Plants, page 24).

Important: Light intensity is more important for the plants than the killifish. This is why you will find more detail on this subject under the descriptions of plants on page 24.

Bottom Material

Killifish show their full colors only against a dark substrate. Seen against a light-colored background, they seem pale and easily frightened. It is also important that the bottom material contain no calcium because most killifish thrive best in soft to moderately hard water (see Water, page 22), and calcium unnecessarily hardens the water.

River gravel and sand (available at pet stores) of various grades of coarseness and as dark as possible makes a good substrate. Gravel with a diameter of up to 3 mm looks very nice. I use a mixture of two-thirds sand (up to 1 mm in diameter) and and one-third fine gravel (up to about 3 mm).

Important: Rinse river gravel or sand before you place it in the tank until the rinse water runs clear.

Crushed basalt and lava also make a good substrate, but they are not suitable for bottom-dwelling killifish (such as *Nothobranchius* and *Pterolebias*). Do not use these materials if you combine killis with bottom-dwelling fish (such as catfish and loaches). Bottom-dwelling fish can be hurt on the sharp edges of the rock fragments.

Planting

Plantings that are suitable for killifish and attractive to the beholder can be achieved with the aquatic plants described on pages 24 and 25. All these plants have relatively low light requirements and are easy to keep in aquarium water that has the properties killis require. Special mention is made of those plants that adapt to extreme water qualities (brackish water, or cool tanks).

Decorative Materials

To create interesting underwater landscapes that also provide hiding places in addition to those offered by plants, you can use the following.

Root wood from bogs (available at pet stores) is wood from tree roots that have been lying in acid soil for years and no longer contain anything that can decompose (available through the AKA).

Willow roots (growing from willow branches that extend into the water) are also excellent and look attractive.

Aerial roots from *Monstera* or *Philodendron* plants can be introduced into the tank directly through a narrow crack (next to the tube leading to the filter). These roots then branch out in the water, forming a fine network, and create a dramatic visual effect.

Important: *Monstera* and *Philodendron* plants quickly deplete the water of plant nutrients. Aquatic plants therefore require frequent feeding (special fertilizers available at pet stores), or they do not grow well.

Suggested planting for a killifish aquarium. 1. Java fern (*Microsorium pteropus*); 2. Java moss (*Vesicularia dubyana*); 3. *Anubias barteri* var. *nana;* 4. Water trumpet (*Cryptocoryne affinis*); 5. *Echinodorus latifolius*; 6. Egyptian lotus (*Nymphaea lotus*); 7. Water pennywort (*Hydrocotyle leucocephala*); 8. *Echinodorus tennelus.*

The Right Aquarium for Killifish

Rocks to be placed in an aquarium must not contain any calcium. This is why igneous rock, such as granite, gabbro, or reddish brown lava (order through a pet supply store), is best. Scrub the rocks thoroughly in water before placing them in the tank.

Setting up the Aquarium in Ten Steps

1. Rinse the tank with cold water.
2. Cover the floor with about $1^1/_4$–2 inches (3–5 cm) of washed bottom material (gravel and sand). If you use heating cables, install them first and then cover with a $1^1/_4$ inch-layer (3 cm) of sand. Top with fine gravel.
3. Slowly fill the tank about one-third with room-temperature water, being careful not to stir up the substrate.
4. Place the decorative materials (rocks and roots) in the tank. Rocks larger than fist size should be placed directly on the tank floor, before you add the sand.
5. Always start planting in the back of the tank and then work toward the front. Do not space the plants too closely; they need room to develop roots and grow.
6. Now fill with water up to $1^1/_4$ inches (3 cm) below the rim.
7. Place the cover, and fill in all cracks with foam.
8. Install the heater (in the tank or the filtration system) and the filter (external or internal filter). Space the filter intake and outflow as far apart as possible.
9. Turn on the filter, heater, and lights, and run the aquarium for two to three days without fish in it. Bottom-dwelling fish are not introduced until two weeks later, when the plants are firmly rooted.
10. Be sure to check the water quality before adding the fish. You will find information on the right kind of water for killifish in the following paragraphs.

Water—The Element in Which Fish Live

Water is another, crucial factor in keeping killifish healthy and in propagating them successfully. It is important that an aquarium hobbyist has a clear understanding of water hardness and water acidity (pH value).

Water Hardness

In nature almost all killifish live in decidedly soft water. Naturally occurring water is called soft if it has a combined hardness of below 2° dH. The combined hardness (dH) is the sum of temporary or carbonate hardness (dCH) and permanent or non-carbonate hardness (dNCH). Any dH value of less than 8 is soft. Water for killifish need not, however, be this soft; in most cases even the rearing water does not have to be this soft (see descriptions, page 53).

Measuring Water Hardness. Both the combined and the carbonate hardness are measured. You can buy easy-to-use solutions for measuring water hardness at aquarium stores. Checking the water hardness once a week is adequate.

The Proper Water Hardness for Killifish. Water with a medium degree of hardness, that is, between 9 and 12° dH, is ideal for keeping and breeding most killifish species. In my case, the Stuttgart city water (10° dH, 6° dCH, pH 7.2–7.5) has proven adequate for killis. Killis can even be kept in water of up to about 15° dH. Make sure the water hardness never drops below 2° dCH (see Dehardening water).

Note: Some killifish species (desert fish and fish from brackish water) thrive best in hard water, and some of these fish need to have sea salt added to their water (see descriptions, page 53).

Dehardening Water: Tap water that is too hard must be dehardened. To accomplish this, you must desalinate the water either completely or—in the case of high carbonate hardness with few other salts—partially. One way is to mix the tap water with distilled (completely desalinated) water. If you need a constant supply of soft water for one or several large tanks, however, you must get an ion exchanger or an osmoregulator, which can remove some or all of the minerals from water. Various types of ion exchangers are manufactured, and the right one for you depends on the composition of your tap water. You can learn more about ion exchangers from books on water chemistry.

Important: Fully or partially desalinated water must always have a hardness of 2° dCH restored; that is, it must be mixed with normal tap water. A carbonate hardness that is too low can result in sudden fluctuations of acidity that can be fatal to the fish.

Water Acidity

The acidity—or alkalinity—of water is expressed in its pH value. A pH of 7 represents neutral water; a pH of less than 7 means that the water is acid; and water with a pH value of more than 7 is alkaline.

Measuring the pH value: There are pH test kits available at pet and aquarium stores. Test kits for the ranges pH 4–7 and pH 6.5–10 are accurate enough. Check the pH value every day!

The right pH for killifish: The best range for killifish is between pH 6 and pH 7.5.

Here, too, desert fish and fish from brackish waters are an exception because they thrive only in alkaline water with a pH value of 7–8 (see descriptions, page 53).

Lowering the pH: Filtering with peat (see Filtration, page 19) works best for this purpose. Peat lowers the carbonate hardness of the water at the same time. So if your tap water is quite soft, merely running it through peat often results in water that is fine, in terms of hardness, for keeping or breeding killifish.

Tip: If you have run an aquarium for some time and have developed some expertise at it, you need not measure the water values except after changing the water. However, if you filter through peat, the water chemistry inside the tank should be checked regularly.

Aquarium Plants

Anubias barteri var. nana
Arum family

Origin: West Africa.
Care: Do not transplant more often than necessary; grows best on roots, cork, or rocks.
Light: Bright or dark.
Water: Up to 15° dH; 68–75° F (20–24° C); pH 6–7.5.
Propagation: Side shoots from the root stock.
Placement: Anywhere in the tank.

Bolbitis heudelottii
Polypody of Fern family

Origin: Africa.
Care: Needs clean water; tie to roots or rocks, and allow the plants to establish themselves there.
Light: Half-shade to bright light; protect against excessive algae growth (*Ampullaria* snails).
Water: Up to 10° dH; 68–79° F (20–26° C); pH 5.8–7.2.
Propagation: Splitting of root stock.
Placement: Background or middle of tank.

Ceratophyllum demersum
Common hornwort

Origin: Worldwide.
Care: Cultivate free-floating; this plant grows fast and keeps losing the oldest parts; cut back periodically.
Light: Bright; use at least one fluorescent tube.
Water: Up to 25° dH; 50–86° F (10–30° C); pH 6–7.5; also brackish water.
Propagation: New shoots.
Placement: Anywhere in the tank; ideal in tanks for rearing fish and in cool water.

Ceratopteris thalictroides
Water sprite

Origin: Asia, Africa, America, and northern Australia.
Care: Submerged, free-standing or floating; leave top of root exposed; needs light.
Light: Bright; use at least two fluorescent tubes.
Water: Up to 15° dH; 75–82° F (24–28° C); pH 6–7.5.
Propagation: Adventitious plants on leaves.
Placement: Middle of the tank or in the back.

Cryptocoryne affinis
Water trumpet

Origin: Malayan peninsula.
Care: Do not transplant unnecessarily; grows well if left alone.
Light: Medium (two fluorescent tubes); undemanding.
Water: Up to 15° dH; 68–82° F (20–28° C); pH 6.5–7.5.
Propagation: Runners.
Location: Middle or background, in clusters.

Echinodorus amazonicus
Amazonas sword plant

Origin: Brazil.
Care: Tanks of 25 gallons (100 L) or more; fine substrate, ideally with bottom heat; needs frequent addition of fresh water.
Light: Medium to intense lights.
Water: Up to 10° dH; 72–82° F (22–28° C); pH 6.5–7.5.
Propagation: Submerged flower stems with adventitious plants.
Location: Plant singly in middle of tank; leave plenty of room.

Aquarium Plants

Echinodorus latifolius

Origin: Central America, northern South America.
Care: Strong lights and a substrate of fine sand with bottom heating; thin regularly, and replace old plants with young ones.
Light: Weak to bright lighting (HQL).
Water: Up to 15° dH; 68–86° F (20–30° C); pH 6–7.5
Propagation: Runners.
Location: Background for flat aquariums.

Hydrocotyle leucocephala
Brazilian water pennywort

Origin: Brazil.
Care: Needs light; otherwise undemanding.
Light: At least two fluorescent light tubes.
Water: Up to 15° dH; 68–82° F (20–28° C); pH 6–7.5
Propagation: Separate the shoots.
Location: Needs clear space; ideal for background and edge planting.

Microsorium pteropus
Java fern

Origin: Southeast Asia.
Care: Adaptable; plant on top of roots or rocks.
Light: Modest requirements; one fluorescent light tube or more.
Water: Up to 15° dH.
Propagation: Adventitious plants on leaves; splitting of root stock.
Location: Anywhere on roots and rocks, especially along the back wall.

Echinodorus latifolius (center)

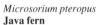

Najas guadelupensis
Pond weed

Origin: Southern United States, Central and South America.
Care: Free-floating, or can be carefully planted; change water and fertilize regularly.
Light: Bright, at least two fluorescent tubes.
Water: Up to 15° dH; 68–86° F (20–30° C); pH 6–7.5; also brackish water.
Propagation: Side shoots.
Location: Allow to float unattached; ideal for rearing tanks.

Nymphaea lotus
Egyptian lotus

Origin: East Africa, Southeast Asia.
Care: Hungry for light and produces underwater leaves only under very bright illumination; otherwise only floating leaves form; fertilize regularly.
Light: Bright (three fluorescent tubes or 1 HQL).
Water: Up to 10° dH; 72–82° F (22–28° C); pH 6–7.5.
Propagation: Runners.
Location: Singly in the middle of the tank or along the edges. Good for providing shade.

Vesicularia dubyana
Java moss

Origin: Southeast Asia.
Care: Requires hardly any care. Floats freely as a bush (spawning plant) or rests on rocks and roots.
Light: Undemanding; moderate to strong light.
Water: Up to 15° dH; 64–86°F (18–30° C); pH 5.8–7.5.
Propagation: Divide the moss cushions.
Location: Anywhere in the aquarium.

Acclimation and Maintenance

Transporting Killifish

The way you transport your fish depends on the length of the trip.

Short Trips

If the trip takes less than two hours, two or more fish can be placed together in a fish-transporting bag (available at pet stores) filled one-fourth full with water. This way enough oxygen is available even for species that require a lot of oxygen. In such a short time the aggression of males toward each other and toward females does not lead to fatalities.

Long Trips

If fish must travel for more than two hours, they should be taken in bags as just described above, but one or at most two fish per bag. To prevent losses through hypothermia or overheating, an insulation bag (available at pet stores) can be inverted over the tranport bag in the winter or summer.

Tip: If you want to ship fish yourself, proceed as follows. Do not feed the fish for one day, so that their bowels are empty. The water in the transport bag should not be burdened with feces because they deplete the oxygen in the water and add organic wastes. The tall, narrow transport bags are filled about one-fifth full with fresh water. This amount of water is quite enough! What is important is that there is plenty of room for air so that oxygen can pass into the water. Do not blow air into the bag from your mouth but always with an aerator. Pack the transport bag into a solid cardboard box, and label the box "Live tropical fish. Please keep at 68–75° F (20–24° C)." If fish are shipped this way there are seldom any losses.

Important Note: At the height of summer and in the winter, when outdoor temperatures can be extreme, the transport bags with the fish in them should be packed in Styrofoam boxes or in fish boxes (available at pet stores) filled with additional Styrofoam pellets.

Quarantine Is Indispensable

If you keep killifish, you need a quarantine tank in which all newly acquired killifish spend a three- to four-week quarantine period before they join other fish in an aquarium. Fish can have parasites even if there are no visible symptoms, and diseases not due to parasites often remain latent until changed environmental conditions (water quality, food, and temperature) lead to an outbreak.

The Quarantine Tank

Set up the quarantine tank exactly as you do the aquarium in which the fish are later to be kept (see The Right Aquarium, page 15). Sterile tanks with a bare floor and devoid of plants and decorations will not do. The quality of the water, too, must be the same as that of the water in the tank where the fish will live. Make sure that no water (not even a single drop) from the quarantine tank gets into the regular tank. Always wash your hands well after doing anything with or in the quarantine tank.

If any disease shows up in the quarantine tank, keep the fish quarantined until the disease has been completely cured. If you detect an incurable disease that is contagious, *all* the fish must be killed (see Killing a Fish, page 43), even those that show no sign of the disease. Do not endanger your entire collection of fish from well-intentioned but ill-guided love for animals.

Caution! I consider it crucial to buy tools—such as a fish net, glass trap, thermometer, cleaning sponge, bucket, and hose—for exclusive use in the quarantine tank so that pathogens are not acciden-

One of the most popular aquarium killifishes,. Aphyosemion gardneri gardneri (steel-blue Aphyosemion), two males and a female.

Acclimation and Maintenance

tally spread with water from that tank into other tanks. If you have tools for general use, you should always disinfect them after they have been in contact with the quarantine tank (see page 26).

Releasing fish into the quarantine tank: This is done as follows:

• Pour about 5 quarts (5 L) of water from the quarantine tank into a bucket.

• Place the plastic bag with the fish inside it into the bucket.

• After 10 minutes, open the bag and gradually add water from the bucket into the bag with a cup until the bag is full.

• Tilt the bag carefully and lift it slightly so that the fish can swim out of it.

• Catch the fish with a fish net and place them in the quarantine tank.

• Replace the water you took from the quarantine tank with fresh water.

• After the quarantine period is over, catch the fish with the net and transfer them to the regular tank.

Important: Make sure that any water the fish are placed in has the proper pH value (see page 23) from the very beginning. Killifish can put up with a change in temperature (up to 9° F 5° C) and even with major deviations in water hardness without ill effects as long as these conditions are temporary.

Cleaning the Aquarium

A well-maintained aquarium is not only an attractive sight in any room but it also offers its occupants optimal living conditions. If you clean your aquarium regularly, you can let it stand for years without having to dismantle it and set it up again all over. Among the routine maintenance chores are the following:

• Change some of the water at regular intervals (see below). Feces and leftover food dirty the water quickly, and even the best filtering system is no substitute for changing the water.

• Clean the filter regularly (see page 30).

• Dead fish and dead plant parts must be removed at once.

• Detritus on the bottom of the tank must be siphoned off regularly whenever water is replaced. (Use a hose for a combination of clear glass and plastic tubing added to a rubber hose; this gives rigidity and makes your siphon more maneuverable. You can control the rate of flow by pinching the hose.)

• Algae on the glass walls or the cover should be removed with a regular household sponge every time before the water is changed.

Note: It is impossible to say exactly how often these chores should be performed. The more fish there are in a tank and the more they eat, the more often the tank must be cleaned—with a lot of fish, about every two weeks. If you have only a few fish and quite a lot of plants in your aquarium, you can let it go four to six weeks.

Changing the Water

Changing the water regularly is important for maintaining living conditions at an even, high quality for the tank inhabitants. Always change only part of the water at any one time, however, so that there are no major fluctuations in water quality (see page 22).

For *community and species tanks* I recommend changing one-fifth of the water every week or one-fourth to one-third every two weeks. In addition to

South American bottom spawners. Above: Two female Pterolebias longipinnis. *Below: A pair of* Cynolebias elongatus.

Acclimation and Maintenance

Introducing fish into a tank. To acclimate a newly acquired killifish to the tank water you have prepared ahead of time, fill the transport bag with the water. Use a cup for pouring.

these routine changes you should change up to three-quarters of the water about every six months. Fish and plants not only give off substances into the water, they also use up crucial elements.

In *breeding and rearing tanks* the water must be changed more frequently. A good general rule is to change about three-quarters of the water every week. Please check information given on rearing specific killifish in the descriptions (page 53).

Important Note: Changing the water regularly prevents killifish suffering a much feared and sometimes deadly ammonia shock. This reaction occurs only if about half the old water is replaced suddenly with new water and the pH consequently shifts from acidic to alkaline. The metabolic wastes that took the relatively harmless form of ammonium when the water was acid turn into toxic ammonia under alkaline conditions (see Poisoning, page 42).

Cleaning the Filter

The larger a filter is in relation to the volume of the tank, the less often it must be cleaned. I cannot therefore give exact rules about how frequently filters should be cleaned. However, be sure not to clean it too early, for most filters only start working really well when they have absorbed some dirt. Filters then sift more finely and run somewhat more slowly.

External Power Filters

Filters of this type should not be cleaned until their efficiency drops noticeably. You will recognize this stage when considerably less water comes through the filter outflow. Now disassemble the power filter, and throw out the filter wadding or rinse it clean with water. The other filter materials and the power filter must also be rinsed with lukewarm water. Reassemble the filter as it was before. The hoses are best cleaned with hose brushes (available at aquarium stores).

Internal Plastic Filters

Filters that use crushed lava do not have to be cleaned until the filter wadding looks quite dirty. Remove the filter from the tank, and take off the cover (sieve) and then the filter wadding. Replace the cover, and fill the filter with water.

After shaking vigorously while covering the sieve with your hand, pour the water out again. Do this several times, and wait until the water comes out clear before you put in the new wadding.

Clean *internal foam filters* by removing the foam rubber, squeezing all the water out of it, and then rinsing it under running water until no more dirt is flushed out.

Caution! Do not use hot water or chemicals when cleaning filters. This destroys the bacteria flora on the filtering substrate. Bacteria are important because they contribute to the filtering process.

Acclimation and Maintenance

Care of Plants

On pages 24 and 25 you will find tips on the care of specific plants that are suitable for tanks with killifish. There are a few basic things you should know to keep aquatic plants from languishing, on the one hand, and from spreading too much, on the other.

Fertilizing: Fertilizers feed plants and stimulate lush growth. Aquarium stores sell fertilizers that are excellent for aquatic plants. It is best to fertilize immediately after every water change and once or twice in between. Maintaining a regular schedule is important. Erratic fertilizing has no beneficial effect and can even interfere with normal plant growth. As a rule, only small amounts of fertilizer should be applied. Keep in mind that the more light plants get, the more nutrients they need. Fast-growing plants, such as hornwort, should also be given plenty of fertilzer.

Important: Do not add fertilizer to a newly set up aquarium until the plants have rooted, usually after about two weeks.

Important: Never use fertilizer meant for house plants. It is deadly poisonous to aquarium fish!

Thinning Plants with Stems: The stems should be shortened when the plants begin to grow horizontally along the water surface. Pinch off the stem with your thumb and index finger in such a way that it ends about 4 inches (10 cm) below the water surface. Remove the lower part, along with the roots, and plant the remaining stem in the bottom.

Thinning Echinodorus: *E. latifolius, E. quadricostatus,* and *E. isthmicus,* all of which are small, fast growing, and very similar to each other, soon send runners through much of the aquarium. You can keep these plants within desired bounds by pinching off the runners. Remove the old, large plants from time to time, and leave some young ones instead. This improves the appearance because old plants do not look as attractive in an aquarium as fresh plants.

Thinning Floating Plants: Floating plants

Plant care. Grasp wilted leaves of sword plants (*Echinodorus*) and water trumpets (*Cryptocoryne*) close to the stem with thumb and index finger and separate from the plant by exerting gentle downward pressure.

should always be kept at about the same density. If they are thinned too much, this leads to great changes in light intensity. These changes especially affect water trumpets (*Cryptocoryne*), which sometimes die off suddenly if the cover of floating plants has been too drastically reduced.

Floating plants can be restricted to a limited area very simply. Attach a transparent air hose with epoxy to two suction cups in such a way that it forms a diagonal line running from the back wall to a side wall. The hose is practically invisible, but it keeps the floating plants from spreading across it.

Transplanting: If plants must be reset, the roots often have to be shortened. I use scissors to cut the roots to a length about equal to the depth of the substrate. The plants then take root quickly.

Transplanting *Cryptocoryne:* With these plants I follow a different procedure. I transplant them only when absolutely necessary, for *Cryptocoryne* not only grow slowly, they also develop new roots very slowly. If the roots are short-

Acclimation and Maintenance

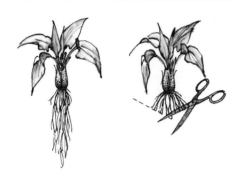

Transplanting. For replanting, the roots must often be shortened. Trim them by about a third or a half.

ened, this unnecessarily slows rerooting. This is why, when I pick up *Cryptocoryne* that have to be transplanted, I wrap the roots around my index finger without breaking them. When the plants are planted in the substrate this way, they take hold quickly and soon sprout new leaves.

I carefully cut off about a third of the roots only with very large plants of this genus, such as *C. usteriana*.

Wilted Leaves: Remove the oldest leaves of *Echinodorus* (sword plants) and *Cryptocoryne* (water trumpets) when they begin to turn brown and soft (wilted). To do this, take hold of the leaf with your index finger and thumb where the stem starts. If you press downward lightly, the leaf separates from the stem. You can also cut off the leaves, but I do not recommend this because the stem stumps that are left are something of an eyesore.

Proper Nutrition

In their natural environment killifish feed mostly on insects that drop onto the water surface and on insect larvae that live in the water. You should make an effort to give the killis in your aquarium live food as often as possible. It is true that most kinds of killifish become used to eating dry food, but dry food should always be given in combination with live food, never as an exclusive diet. A few killifish species refuse dry food altogether and have to be given only live food if they are to stay healthy. When this is true, the fact is noted under the individual species in the descriptions (page 53), where you will find exact instructions on composing different diets.

Live Food

The first thing to be aware of is that live food (which, for the purposes of this discussion, means actual, living food animals as well as fresh food animals that have been subjected to minimal processing to preserve their nutritional value) comes in three different forms:
• fresh (live food animals)
• frozen
• freeze-dried

Fresh Live Food

Food animals that are alive most closely resemble the food killifish eat in nature and are therefore the most desirable element in the diet of killis kept in an aquarium. The movement of the prey makes the fish chase after it and try to catch the animals. This, of course, has a highly positive effect on the physical condition of the fish, which suffer from lack of exercise in the aquarium. Live food animals also contain all the important nutrients fish need, as well as a high proportion of roughage (which stimulates digestive activity) and many vitamins and trace elements.

Where to Get Live Food Animals

There are several ways to obtain suitable live food animals for killis.

Buying the animals from pet dealers saves you a lot of time, and if you purchase them from a well-run pet supply store you need not worry that they might introduce pathogens or parasites into your aquarium. Unfortunately, many food animals of high nutritional value (see page 34) are not sold year-round. However, it always pays to ask. *Tubifex* worms and white mosquito larvae are usually the only kinds of live food available all year.

Raising suitable food animals yourself (see page 38) makes you less dependent on commercial offerings, but you will have to spend more time on your hobby.

Catching animals yourself: Food animals for your fish should always come from clean waters in which no fish are living. Live food from fish ponds may introduce pathogens and parasites, such as fish leeches and fish lice, into your tank. Make sure no *Hydra* or eddy worms (planarians) get into a tank with spawn or fry in it, for both these small organisms are predatory (see drawing, page 36). Grown fish, on the other hand, are not harmed by *Hydra* and planarians.

Feeding. Live food animals are an essential part of a healthy diet for killifish living in an aquarium.

Proper Nutrition

Caution: In some places state and federal environmental protection legislation prohibits removing flora or fauna from public waters. Find out what regulations apply to the American killifish species in question by inquiring at appropriate local authorities. For privately owned bodies of water, you must ask the landowner's permission, of course.

Suitable Food Animals

The range of food animals suitable for killifish is huge, and to name them all is beyond the scope of this book. Here I describe the most common and most nutritionally valuable food animals that cannot be raised at home. I also include some tried and tested tips for the storage and feeding to fish of food animals bought at the pet store or collected in the wild. You will find more complete information in some of the literature listed in the back of the book (see Addresses and Literature, page 69).

Black mosquito larvae (Culex, biting mosquito larvae) live from early spring to the late fall in pools and puddles of water just below the surface. Catch only as many as you can feed to your fish within four to five days, or freeze the rest (see Frozen Live food, page 35). Keep the larvae in cold water in a small container with a cover. Before removing the cover to take out some larvae, quickly turn the jar upside down over the bathtub. This way the hatched mosquitos drown.

White mosquito larvae (Corethra, phantom midge larvae) are clear as glass. They hover horizontally in clear, still water. Especially at night, large numbers of them are directly under the water surface and can be caught easily. Place them on damp newpapers, wrap in more newspapers, and keep in the refrigerator (in the butter compartment). You can also keep them in tubs of cold water. Do not panic if some of the larvae hatch; the mosquitoes do not bite. If you should come upon large amounts of the larvae, you can freeze some of them (see Frozen Live Food, page 35). Unfortunately, many small killifish species refuse to eat this highly valuable food for more than two weeks at a time.

Important Note: Be careful if there are baby fish in the tank! White mosquito larvae are predatory and easily subdue fry up to their own size. Refrain from using the larvae for food in such situations.

Red mosquito larvae (Chironomus, larvae of nonbiting midges) live in the bottom mud of waters with high levels of organic matter and are therefore hard to catch. They are sometimes found in considerable numbers in rain barrels, however These larvae should not be fed exclusively for more than three to five days, especially to small killifish species. If they come from unclean water they often contain heavy metals or organic products of decomposition, which can lead to health problems in fish fed too exclusively with these larvae.

Water fleas, such as Daphnia and Cyclops, live in clean, still water. Both can be fed exclusively for extended periods. They also provide roughage, which is useful if you frequently feed your killis red mosquito larvae, Tubifex worms, or beef heart (see page 35). Water fleas are difficult to keep alive for any extended period, but they freeze well. They can be caught in great numbers in the summer.

Cyclops can be kept in a tub of cold water. Small killifish species and young fish are especially fond of this food.

Important: Give Cyclops only to fry large enough to cope with them (see Food for Young Fry, page 37). If the fish are too small, they are eaten by the predatory Cyclops.

Tubifex worms are especially useful for "everhungry" killifish species and for fish that grow rapidly and have a highly active metabolism (see descriptions, page 53). These red worms live in the mud of dirty water, especially in the drainage ditches of waste treatment plants. The worms form clumps that can be picked up by hand or with a ladle. Before using Tubifex worms as live food, place them in slowly running water for two or three days until their muddy intestinal contents have been washed away. Once a day, carefully lift up the cluster of worms and rinse away the dirt that has collected underneath. Tubifex worms die in stagnant water.

Feed these worms in small amounts only; otherwise they quickly burrow into the bottom. The quality of the water in the aquarium deteriorates considerably if you drop a large number of Tubifex into it. It is best to shake a small ball of worms with water and then distribute the brew over the surface of the water. This is the most effective method to ensure that the fish will eat most of the worms before they can sink to the bottom. These worms are also available dried, freeze-dried, and pelletured. They are suitable as a food supplement only, as they neither are very rich in vitamins nor provide sufficient roughage. Tubifex worms should *not* be frozen.

Earthworms are eaten eagerly by large killis, as you will observe if you have such fish. Depending on their size, the worms should be cut into pieces before being given to the fish.

Ants also form a large part of the diet of killifish in the wild. For aquarists, red garden ants are the easiest to gather. Eating red ants brings out the reds in your killis with special vividness. Do not collect other kinds of ants because many of them are endangered species that are therefore protected.

Frozen Live Food

The nutritional value of frozen food is almost as high as that of fresh foods. Frozen food is easy to store and to use. You can either buy it at a pet store or freeze it yourself. If at some point you catch a lot of live food animals or have bought more than you can feed immediately, freeze what you can not use right away to have on hand later.

Live food for fish is frozen the same way as food for your own consumption. It is important that the freezer be fully cold so that the food freezes immediately. Frozen food kept at slightly below 0° F (−20° C) keeps its full nutritional value for one to two years.

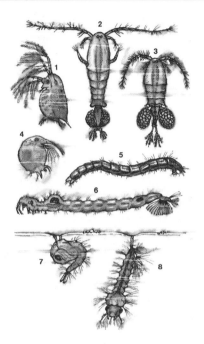

Food animals. The healthiest diet for your killifish is biologically clean live food. You can either catch the animals depicted here yourself or buy them at a pet or aquarium store.
1. *Daphnia* (water flea) 2. Female *Diaptomus* (water flea) with egg sacs. 3. Female *Cyclops* (water flea) with egg sacs. 4. *Bosmina* (water flea) 5. Red mosquito larva (*Chironomos*). 6. White mosquito larva (*Corethra*) 7. Pupa and 8. larva of biting mosquito (*Culex*; black mosquito larva).

Food animals that are suitable for freezing include Daphnia, Cyclops, all kinds of mosquito larvae, and freshwater amphipods. At pet stores you can also get brine shrimp (*Artemia salina*) and, for larger killifish, opossum shrimp (*Mysis*), krill, and sand shrimp (only for very large killifish).

Beef heart (raw) is also recommended for killifish, but it should be fed neither alone nor very often. First remove all the fat and sinewy tissue;

then cut into strips and freeze individual portions. Grate the frozen strips with a regular kitchen grater before feeding beef heart to the fish.

Tip: You can add vitamin drops (see Vitamins and Trace Elements, page 36) to grated and thawed beef heart. This stimulates enormously the growth of the fish.

Mussel meat is also a highly nutritious food for killifish, and it is easily frozen. Cook the mussels in saltwater until they open. Mussels that remain shut are not usable. Then remove the meat from the shells and freeze it in individual portions. Before feeding it to the fish, proceed as with beef heart. Other shellfish can be used, but they are more expensive without being any better.

Important: Be sure to thaw frozen food completely before giving it to the fish. Food that is too cold can cause intestinal upsets in killis. After thawing the food, rinse it under running water in a sieve to prevent juices produced by the freezing from getting into the tank water.

Freeze-dried Live Food

Freeze-dried live food sold by pet stores consists mainly of freeze-dried Daphnia, Artemia, and red mosquito larvae, and sometimes a few other kinds. Food in this form is light and floats on top of the water for a long time and thus is available primarily to hungry surface-dwelling killis (see descriptions, page 53). I consider a diet made up exclusively of freeze-dried food too one-sided if used over long periods.

Dry Food

Pet stores sell dry fish food in the form of flakes, tablets, and granules. Although this food contains many important nutrients, vitamins, and trace elements, as well as roughage, you should feed it to your killis only in combination with live food.

Food flakes often lead people to overfeed, so that the fish become too fat. To prevent this danger, follow the feeding rules (see page 37).

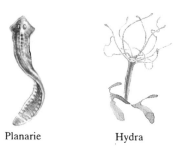

Planarie   Hydra

Potential enemies in a rearing tank. Eddy worms (*Planaria*) and *Hydras* are predatory and may destroy a great many fish eggs or fry in a tank.

Food tablets are pressed against the aquarium glass. It has been my experience that although some species become used to this form of feeding, they do not really like it. In accordance with their natural feeding behavior, killis like to snap at their food rather than "graze." This is why I consider this form of feeding inappropriate.

Food granules are particularly useful for the larger species. For smaller killis, the granules should be soaked in water before being fed to the fish.

Vitamins and Trace Elements

You must add these supplements only if you feed your fish exclusively and for extended periods on beef heart, Tubifex worms, and dry food (which loses much of its vitamin content if stored too long).
• Multivitamin drops (available at drug stores and pet shops) are added to thawed beef heart or to dry food just before feeding. If vitamins are given with Tubifex, they should be added a few hours before the feeding.

• B complex vitamins (available at drug stores) are especially salutary for fast-growing killifish species.
• Vitamin D is available at drug and pet stores. If mixed with some fresh lemon juice, it prevents skeletal malformation.
• Trace elements (available at pet stores in powder form designed for terrarium animals) are an important supplement, especially to dry food and beef heart.

Food for Young Fry

Live food animals are a must if you want to raise killifish successfully. In my experience dry food and infusorians have proved inadequate for raising these fish. Almost all killifish eat Artemia nauplii (see page 39) immediately after hatching. The few species whose fry need more microscopic food for the first few days (see descriptions, page 53) can first be given vinegar eelworms and microworms (see page 39). After a few weeks, these young fish, too, eat freshly hatched Artemia nauplii. Larger species and somewhat older fry can be fed Cyclops and Daphnia, as well as grindal worms (see page 38).

A Good Rule of Thumb: Give young fry only food that is no larger than their eyes.

Feeding of Breeding Stock

The fish that you have selected for breeding should be fed exclusively and plentifully on live food. Specific feeding suggestions for different species and subspecies are given under the various killifish genera in the descriptions (starting on page 53). Dry food should not be used at all in breeding tanks and is totally inadequate to stimulate the spawning.

Feeding Rules

1. Feed adult killifish at least twice a day! In nature, they eat all day long.

2. Young fry must be fed *several times* a day.

3. Only give as much food as the fish will eat within about 10 minutes!

4. Institute one fasting day a week for fully grown fish!

5. Best feeding times: in the morning, about 15 minutes after turning on the lights, and again in the late afternoon, several hours before turning off the lights.

6. Turn off the filter when you feed the fish so that the food is not sucked into the filter. If you use interior foam filters this is not necessary.

7. Food that is not consumed by the fish must be siphoned off with the help of a tube. Reduce the amount of food the next time.

8. *Ampullaria* snails and bottom-dwelling fish are useful in a communal aquarium for disposing of leftover food. With this kind of tank, you should feed enough so that a little of the food drifts to the bottom.

Raising Food Animals

There are some nutritionally valuable food animals that you can raise yourself without investing too much time or money. You can buy breeding colonies and breeding containers through pet dealers, aquarists' associations, and specialized magazines. Breeding containers can also be obtained from laboratory suppliers.

Fruit Flies (*Drosophila*)

These flies can easily be cultivated in large numbers. For aquarists the flightless laboratory strains are best, such as the small, stumpy-winged *D. melanogaster* (vestigial form) and the larger, flightless *D. hydeii*.

Culture Containers: Glass canning jars are often recommended for this purpose, but I think they are neither practical nor hygienic. I advise you to purchase laboratory flasks. These are made of plexiglass and come complete with foam rubber stoppers.

Culture Medium: To feed the grubs prepare a mush as follows:

Ingredients: one 10-ounce can of plums, one half-pound package of oatmeal, one orange (with the peel), and one banana.

Puree all the ingredients in a mixer; add a cup of vinegar and enough water to get a runny but not watery consistency. Place about 1 inches (2–3 cm) of this in the bottom of the container, and sprinkle with a little dry yeast. Place some crumpled paper on top for the flies to crawl up on.

Starting a colony: 20–30 flies per jar are enough. The smaller species develops in 12–20 days, depending on the temperature; the larger ones takes three to four weeks.

Removal: When you are ready to remove flies to feed to the fish or as colonies for further breeding, place the jar in the refrigerator for a short time. The cold makes the flies sluggish and unable to move so that you can take them out of the jar without diffi-culty. Unfortunately, fully developed flies live only a few days and therefore must be used as food or for breeding immediately.

Tip: To prevent mold from forming, it is often recommended that the drug Nipagin be added to the culture medium. However, this drug is suspected of causing hereditary damage in fish. It is therefore better to do without it; orange peel also prevents mold.

White Worms and Grindal Worms

Both white worms (*Enchytraeus albidus*) and grindal worms (*E. buchholtzi*) are related to *Tubifex* worms and to earthworms. The somewhat larger white worms are a good supplementary food for fish measuring over 2 inches (5 cm); for smaller species and fish that are still growing, grindal worms are more suitable.

Culture Containers: Fill plastic containers about 2 inches (5 cm) deep with damp peat that has been soaked in water for several days and then squeezed until quite dry and rubbed to a fine, crumbly consistency. Add the breeding colony to this. The containers should have tiny air holes and be kept in a dark place.

White worms reproduce at 50–59° F (10–15° C). Feed them oat flour and dry milk in a proportion of 1:5.

Grindal worms reproduce at 68–77° F (20–25° C). Give them oat flour and powdered milk in a proportion of 1:5.

The ingredients are mixed and spread over the loosened and dampened peat surface. Then spray some water over the top.

Caution: If not enough air enters, the food begins to ferment and heats up so much that the worms will crawl out of the containers. Sometimes mold also forms. If mites begin to grow in the colony, the worms must be washed and started again with fresh peat. Breeding colonies should be replaced with new ones after two months at most.

Removal: Press a piece of glass gently against the top of the peat. The worms will collect on the glass. Pick up the glass and rinse off the worms.

Vinegar Eelworms

These small food animals are worms, too, and are especially suitable for raising tiny fry still too small to eat Artemia nauplii.

Culture Medium: Vinegar eelworms live in diluted wine vinegar.

Ingredients: two parts wine vinegar, one part water, and one pinch sugar.

Culture Container: Fill a bottle, such as an old milk bottle, half full with the vinegar solution. Then pour in half a cup of liquid from an existing culture (for sources of cultures, see Addresses, page 69), and place the bottle in a dark, cool place. The eelworms will multiply quickly.

Removal: When you want to take out eelworms to feed to the fish, you must pour some of the liquid in the bottle through a coffee filter. Catch the strained liquid, and return it to the bottle. Rinse the worms in the coffee filter briefly under the tap, and then pour them into the rearing tank.

For miniscule fry, take an air hose and siphon up liquid from the bottom of the culture, pour the liquid through a filter, and add the filtered liquid—which contains microscopic young eelworms—to the tank drop by drop. These infinitesimal worms are eaten by even the tiniest fry and represent the only way to rear the fry of some species.

Artemia Nauplii

The larvae (nauplii) of brine shrimp (*Artemia salina*) are the food most commonly used for rear-ing fry. Artemia eggs are easy to store and are thus always available for the cultivation of live food. Many pet stores sell live brine shrimp, and the eggs are commercially available in vacuum packed cans.

Take only what you need from the can, and place this portion in a 1–2% household salt or seawater solution at a temperature of 74–80°F (24°C). The eggs will hatch within 24–36 hours.

Before using them as food, pour them through a strainer. Be sure that you buy only eggs that are guaranteed to hatch (check the package) and that the can is moisture free. After removing the portion you need, close the can tightly, tape the cover closed, and store the can in a cool, dry place. If you follow this advice, you may find the brine shrimp eggs still viable after 10 years!

Do not make brine shrimp the exclusive food for your fish, because in time the high salt content will be harmful to them. A one-item diet like this lacks variety and balance, and the fish will soon show signs of deficiency. As growth food for small fry and as a stimulant for adult spawning, however, brine shrimp are without equals.

Microworms

These worms (*Anguillula silusiae*), which are threadlike organisms about 1/8 inch (2 mm) long, are fed to fry that may be too small to cope with *Artemia* nauplii.

Culture Medium: Runny oatmeal made with heavily diluted milk.

Culture Container: Place about 3/4 inch (2 cm) of the culture medium in small glass jars that can be closed tightly, and add a small amount of an existing culture to each jar.

Removal: After a very few days the microworms start climbing up the glass walls and can easily be taken out with your finger or a soft brush.

Diseases of Killifish

Preventing Disease Is Better than Curing It

On the whole, killifish are not very susceptible to illness. There are a few diseases that could be called typical for this fish family, however. Most of them are caused by viruses and bacteria and by plant and animal parasites. Fish are most likely to get sick if their organism has been weakened through stress and unfavorable living conditions.

General measures for preventing health problems:
• Do not overcrowd your aquarium; otherwise the fish suffer from stress.
• If you set up a community aquarium, combine only nonaggressive species that have identical or similar requirements.
• It is important to feed the right kind of food in the right amounts.
• Change the water regularly.
• Check and correct the pH value, hardness, and temperature of the water in the aquarium regularly.

Specific measures:

Get into the habit of always observing certain rules when doing anything in the aquarium or handling fish and equipment used in connection with your hobby.

Dead fish must be removed immediately!

Aquarium water not from your own tanks should be regarded as a possible source of infection. Make sure your fish have no contact with such water. The same caution applies to aquatic plants and other aquarium dwellers, like snails and shrimp.

Newly acquired fish, snails, shrimp, and aquatic plants should always undergo three to four weeks of quarantine (see The Quarantine Tank, page 26).

New decorative materials (roots and rocks) from tanks other than your own must always be disinfected before being introduced. Use a solution of 1 heaping teaspoon potassium aluminum sulfate (available at drug stores) disolved in 1 quart (1 L) of water. I soak all objects of decoration in this solution for about 24 hours.

Fish nets and other tools should be placed in a bucket with the same disinfecting solution after use (submerge up to the handle!).

Tanks must be disinfected if a large number or all the fish in them have been sick and they are to receive new fish. Smaller tanks are filled with the disinfecting solution already discussed. Larger tanks should be wiped out thoroughly several times with the same solution but mixed twice as strong or stronger.

Important: After being disinfected, all objects must be well rinsed.

The Most Common Diseases of Killifish

Most of the diseases described here can be successfully treated with commercially available medication. If you have to resort to these, be sure to follow the instructions that come with them. In some cases I suggest specific drugs I have used with good results. Directions for the use of these drugs are given on page 43. Remember that all drugs must be stored out of the reach of children.

White Spot Disease
Signs of illness: Small, white bumps (up to 1 mm in diameter) that look like grains of cream of wheat appear on the skin. The fish rubs against objects and pushes off with a jerky motion. Protuberant gills and accelerated, heavy breathing are also signs of this disease.

Cause: Ciliate protozoans (*Ichthyophthirius*)

that usually found their way into the tank with newly acquired fish.

Treatment: Commercially available medications (follow directions) or chlortetracycline.

Velvet Disease

Signs of Illness: Velvetlike, yellowish to brownish coating on skin; sometimes accompanied by reddening of skin. This disease is particularly common in killis of the genus *Nothobranchius*.

Cause: Flagellate protozoans (Flagellata), *Oodinium pillularis*, and other protozoans.

Treatment: The same as for white spot disease.

Cloudy Skin

Signs of Illness: Milky, brownish white film on skin and also on gills; in a massive infestation the whole body is covered by a grayish coating. The condition is hard to spot in young fish.

Cause: Flagellate protozoans (Flagellata); ciliate protozoans.

Treatment: Bathing in potassium permanganate (see page 43) or in a solution of table salt. The pathogen dies in temperatures above 86° F (30° C). Treat fry and adult fish that are heavily infested with chlortetracycline. Outbreak of the disease in young fry can be prevented (see Rearing, page 51).

Fungus Diseases (Mycoses)

Signs of Illness: Grayish to white growths from the skin resembling tufts of cotton wool. These growths commonly occur on the skin and the fins but can also attack the eyes, the gills, and the mouth (especially in large, old *Pterolebias*).

Cause: Fungi of the genera *Saprolegnia* and *Achlya*. They establish themselves mostly on skin damaged by injuries or bite wounds. A fungus infection can also be triggered if the fish are too cold.

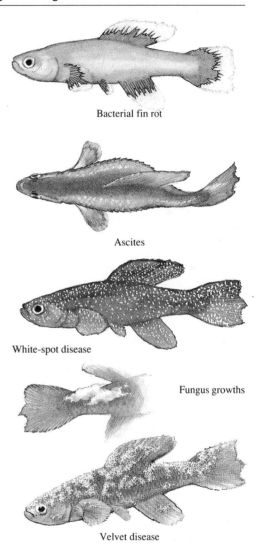

Bacterial fin rot

Ascites

White-spot disease

Fungus growths

Velvet disease

Fish diseases. These drawings depict the most common diseases of killifishes. Fish diseases are caused by viruses, bacteria, and plant and animal parasites. Fish debilitated by unfavorable environmental conditions or stress in the aquarium are most susceptible to disease.

Diseases of Killifish

Treatment: In many cases, raising the temperature or moving the fish into fresh, clean water with some salt added (1 teaspoon per 25 gallons (100 L) takes care of the problem. Treat heavy infestations with commercially available medications, following the instructions, or with potassium permanganate. If nothing else helps, use chlortetracycline.

Fin Rot

Signs of Illness: Frayed fins, often inflamed, discolored, and slimy so that the rays are stuck together. Swellings at the base of the fins and progressive rot until the tail fin drops off. Early stage: clouding of fin rims.

Cause: Various bacteria, especially of the genus *Pseudomonas*.

Treatment: Use commercially available medications (follow instructions) or chlortetracycline. As a deterrent keep the water quality high.

Thyroid Problems

Signs of Disease: Swelling in the throat area often leads to obstruction of the mouth at an advanced stage.

Cause: Malfunctioning thyroid gland.

Treatment: In most cases treatment with potassium iodide with added iodine (see page 43) clears up the problem.

Worms

Signs of Illness: Inflamed, prolapsed anus with a white or red "thread" hanging fom it; emaciation.

Cause: In killifish, usually threadworms (nematodes).

Treatment: Difficult; try commercially available medications first, following directions. It is better to kill heavily infested fish whose organisms are already seriously weakened. Try to prevent infestations by looking fish over carefully before buying them, and then quarantine them. Quarantine an aquarium that has contained worms, and apply the strictest rules of hygiene to prevent spreading of the parasites.

Ascites, Protruding Scales, and Tuberculosis

Signs of Illness: Tumors, bloated body, pallor, protuberant scales, blisters, bulging eyes, and lurching are just a few of the symptoms.

Causes: Bacteria, viruses, and fungi.

Treatment: There is no treatment that is likely to cure these conditions, and it is therefore better to kill afflicted fish. Try to prevent these diseases by providing good care and keeping the tank in optimal condition; a one-sided and overly rich diet is particularly likely to weaken the organism and make it susceptible to disease.

Poisoning

If your killifish gasp for air, careen, move sluggishly, dart around, or behave abnormally in other ways, they may be suffering from poisoning. You can often save the fish if you act immediately and change some of the tank water or catch and transfer the fish to a tank with uncontaminated water of appropriate quality. To prevent poisoning you have to know what substances can be toxic to your fish.

Metals, like zinc or copper, either singly or in combination, can lead to severe metal poisoning if they find their way into a tank.

Note: Herbicides that kill algae often contain copper sulfate. Do not use these chemicals!

Organic waste products, like ammonia, nitrite, and nitrate, can reach harmful proportions if the water is not changed regularly enough or if an aquarium is overcrowded.

Chlorine damages the mucous membranes and the gills of fish. Allow heavily chlorinated water to stand for a day, or use a water purifier (available at aquarium stores).

Diseases of Killifish

Changes in behavior	Possible causes of illness	Treatment
Fish dart around rapidly in water	Poor water quality (poisoning); parasite infection	Check water quality and technical apparatus; place fish in impeccable water or change some of the water (see page 29) if fish have parasites, initiate treatment
Gasping for air at water surface	Insufficient oxygen due to malfunctioning of apparatus or overcrowding; poisoning	Change some of the water immediately, or place fish in impeccable water (see page 29)
Fish lie motionless on bottom, unable to swim; some of the fish are exceptionally shy, have torn fins, or lack some scales	Old age; being too cold Wrong combination of fishes	Kill old fish (below); check water temperature (heater); (see page 16)
Fish do not eat	Serious warning sign: wrong diet; early symptoms of a disease; older annual fish hardly eat when close to dying	Provide more varied diet (see page 33); check fish for signs of diseases

Directions for Using Drugs

All the drugs discussed here are available at pet stores or at drug stores.

Chlortetracycline: This antibiotic (for which you need a prescription) is also sold under the name of Aureomycin. The dosage is 15 mg per 1 L of water. Dissolve in the water before adding to the aquarium. Adding 1 tablespoon of table salt to every 2.5 gallons (10 L) of tank water enhances the effectiveness of the treatment. Change one-fourth of the tank water after a week. At this strength the drug can be used even on tiny fry.

Iodine in Potassium Iodide: Dissolve 1 g iodine and 100 g potassium iodide in 1 L of water. Use 0.5 ml of the liquid per liter of tank water as a permanent additive.

Salt Solution: Use 10-20 g table salt per 1 L of water. Bathe the fish in this solution for about 20 minutes.

Potassium permanganate: The dosage is 1 g potassium permanganate per 25 gallons (100 L) water. Bathe the fish in this solution for about 30 minutes.

Important: If you resort to these treatments, be sure always to observe the concentrations indicated. When treating conditions caused by parasites (white spot, velvet disease, skin clouding, mycoses, and fin rot), always try commercial medications first. Do not use chlortetracycline unless the pathogens have developed a resistance to the commercially available drugs.

Killing a Fish

As I have mentioned several times, it is sometimes more humane to kill fish that are very sick than to let them linger. If you have to do this, sever the spinal cord just below the head with a very sharp knife or with sharp scissors. By doing this you spare the animal unnecessary suffering.

Breeding Killifish

For most aquarists, success in rearing offspring of their aquarium fish is the high point of their hobby. They are justly proud of this accomplishment since it proves that they have provided their fish with the correct care and environment—the conditions necessary for reproduction.

Which Fish Are Suitable for Breeding?

When you select fish for breeding, there are several things you should pay special attention to.

No Physical Defects: Fish that are to be bred should be of perfect external appearance. No fish with a curved spine, a shortened or too thin caudal peduncle, or with partially missing fins should be bred.

The Right Age: According to my experience, plant spawners (see page 47) are at their height of fertility in their second year; *Diapteron* species, in particular, do not spawn really well until then. Bottom spawners (see page 48) spawn best when half-grown and before they reach their full size. Large *Roloffia* species are a good example: when they are about 2 inches (5 cm) long, they produce plenty of eggs, but once they have grown beyond 2 3/4 or 3 1/4 inches (7 or 8 cm) their fertility diminishes noticeably. Fully grown or old killifish are often hardly able to reproduce at all, and very young ones are not very productive either.

Characteristic Coloration: The typical markings of a species, such as stripes and dots on the body and bands along the edges of the fins, should be bright and well defined. Missing stripes or stripes with breaks in them are quite common in almost all striped *Aphyosemion* species. This is not a serious defect, but it is not desirable.

Typical size: Check to make sure your breeding fish are approximately the size that is typical of the species in nature. Fish that have been bred in aquariums are often larger and heavier than the original strain in the wild. This phenomenon is called acceleration.

The Breeding Tank

What a breeding tank should look like depends on:
• the size of the killis you want to breed,
• the spawning habits of the fish (plant spawners, see page 47; bottom spawners, see page 48),
• and the method of breeding.

Size: A few basic tips:
• For killifish species up to 2 inches (5 cm) long I use tanks holding 2 quarts (2 L) or more for a limited breeding period (see Limited-term breeding, page 49) and tanks holding 2.5 gallons (10 L) or more for long-term breeding (see page 49).
• Since, except in the case of lamp-eyes, these small killifish are always bred with only one pair or a trio per tank, tanks with a capacity over 2.5 gallons (10 L) make sense only for very aggressive species.
• I breed killis up to a size of 4 inches (10 cm) in tanks holding 5–8 gallons (20-30 L). This size is adequate for both limited-term and long-term breeding.
• I place all killifish above 4 inches (10 cm) in length in a 9-gallon (35-L) tank for limited-term breeding. If I use the long-term method, I place the fish into tanks of 13–26 gallons (50–100 L).

Important: Do not forget to place a glass cover on the breeding tank!

Setting up the Tank: The only objects needed are something for depositing the eggs on (see Spawning Aids, page 47) and hiding places for the females (Java moss, Java fern, pond weed, rocks, and root wood from bogs).

Killifish of the genus Aphyosemion. Above: Aphyosemion australe (lyretail or lyretailed Panchax,), two males and a female. Below: A male Aphyosemion filamentosum red.

Breeding Killifish

Water: The correct water quality for breeding is given in the descriptions of the species starting on page 53. Always make sure that the water is clean and fresh and contains as little nitrate as possible. The best way to measure the nitrate is with a test strip (available at aquarium stores or from laboratory supply firms). In the case of long-term breeding, up to two thirds of the water must be changed every week. If you do limited-term breeding, there is no need to change the water since new water is used each time the fish are bred.

Location: A breeding tank should be placed in dim light. Ideally it should be briefly exposed to the morning sun. This is the time most killifish species prefer for spawning.

Spawning Aids for Plant Spawners

In nature, plant spawners attach their eggs to aquatic plants or their roots. In a breeding aquarium you can offer them the following objects to spawn on.

A *mop of synthetic yarn* (for instance, acrylic knitting yarn in brown, gray, green, or black colors) works very well. The eggs are easy to see on the yarn, the mop is easy to clean (boil it!), and no harmful substances are given off into the water. Natural fibers are not desirable because they decompose in water. On page 48 there is a drawing of such a mop.

How to Use the Mop: Soak the spawning mop for about 15 minutes in hot water before use; then rinse to remove any traces of chemicals and dyes that may be present. If you attach the mop to a cork, it will float in the water. For killifish species that prefer to spawn closer to the bottom (see descriptions, page 53), remove the cork so that the mop can sink to the bottom.

Peat fibers (available as "filter peat" in aquarium stores) must first be soaked in a separate container until they no longer float on the surface (do not boil them!). After soaking the peat, squeeze the water out of it and place it in the breeding tank.

Cork pieces and pumice stones (both float in the water) are spawning substrates popular with lamp-eyes.

Cork pieces consisting of rough natural cork with deep fissures offer an ideal site for these killifish to spawn on. Place squares (4×4 inches or 10×10 cm) of cork in the breeding tank with the cracks toward the bottom.

File grooves $^1/_{16}$–$^1/_8$ inch (1–2 mm) wide and deep into a pumice stone before placing it in the tank. Lamp-eyes like to deposit their eggs in these grooves.

Note: Cork and pumice stones are also useful in a community aquarium that includes a group of lamp-eyes. The cracks and grooves offer the eggs relative safety from other fish that like to eat fish eggs. Once the spawning has taken place, the cork square or pumice stone must be transferred to a rearing tank.

I use *stainless steel grates* for killifish of the genera *Cyprinodon, Aphanius,* and *Fundulus* (see descriptions, page 53). I do this because these fish prey on their own spawn. The grate must cover the entire floor of the tank and be about $^3/_4$ inch (2 cm) above it. Pour enough sand over the grate to cover it. Instead of adding sand you can place green perlon wadding (available at aquarium stores) underneath the grate. Small pieces of steel grate are often sold as "scrap metal" by hardware stores.

Tip: If the fish eggs are to be stored wet (see Handling the Eggs, page 49), you should use a mop, cork piece, pumice stone, or steel grates. If the eggs are to be stored dry, peat fiber is more practical.

Members of the genus Nothobranchius *are very brightly colored.* Here is a pair of *Nothobranchius melanospilus* (Beira nothombranch).

Breeding Killifish

Spawning mop for plant spawners. Tap two long nails partway into a board about 12 inches (30 cm) apart, and wind yarn around them. Tie the yarn tightly together in the middle with a short string, and then cut the yarn at both nails. Tie a bottle cork below the knot so that the mop floats on the surface.

Spawning Aids for Bottom Spawners

Bottom spawners deposit their eggs either on the bottom of the water (bottom spawners) or dive into the bottom mud to spawn there (substrate divers). In breeding tanks the following spawning aids have proven useful.

Peat is the most commonly used spawning substrate for these fish. Peat from the high-lying bogs is best. I use the "white" kind of this peat, which I mix with one-third of similar but "black" and not too far decomposed peat (both are available at aquarium stores or garden supply centers). This peat is then soaked until it no longer floats (no need to boil it!).

To keep better track of the eggs, I mix the peat just described with one-third peat fiber. The fibers are first cut into pieces $1\frac{1}{4}$–2 inches (3–5 cm) long. In this mixture the eggs are much easier to spot than in pure peat. Squeeze the peat in a net, and place it in a spawning container for bottom spawners.

Spawning Container for Bottom Spawners: So that you need not cover the entire bottom of the tank with peat, you should place a separate spawning container into the breeding tank (see drawing, page 49). I recommend two kinds of containers, both of which have worked extremely well for me. The first are freezer containers made of soft plastic, with covers. These containers should be at least 4 inches (10 cm) tall. Cut a hole in one side a little below the cover. The hole must be big enough for the fish to swim through it comfortably. The cover keeps the peat in the container. The other type of container I like is empty plastic beverage bottles. Sizes from 8 ounces to 3 quarts (0.5–3 L) are excellent for substrate-diving annual killis of all sizes. Bottom spawners also accept these bottles for spawning. Spawning containers for bottom spawners must have a layer of peat about $\frac{3}{8}$–$\frac{3}{4}$ inch (1–2 cm) deep at the bottom; substrate divers need peat up to 6 inches (15 cm) deep (depending on the size of the species). If you use this type of container you can easily remove the peat along with the eggs and drain it in a fine sieve (see Handling of the Eggs, page 49).

Note: For bottom spawners you can also use shallow dishes (freezer containers 2 inches or 5 cm, tall). Place them on the bottom of the breeding tank. This keeps the spawning peat from accumulating too much detritus from leftover food and feces.

Breeding Methods

There are two methods of breeding killis:

Limited-term Breeding: In this method, a pair or a trio (one male, and two females) of breeding

48

Breeding Killifish

Breeding tank for bottom spawners. Use a clear plastic bottle cut off at the neck as a spawning container for bottom spawners. For killifish that dive into the substrate to spawn, the tank must be tall enough for them to dive from above into the bottle containing the peat.

fish are placed in the tank for 12–24 hours. After they have spawned they are removed from the breeding tank.

Before you place the fish in the breeding tank, you must keep them segregated by sex until the females show that they are ready to spawn. In many species, the eggs are easy to detect inside the body. They show up yellowish to orange through the skin in front of or above the anal fin in a female about to spawn.

Limited-term breeding is especially recommended for species aggressive toward their own kind (see descriptions, page 53) and for those that tend to eat their own eggs. These cannibalistic killis include many small *Rivulus, Cyprinodon,* and *Aphanius* species. Tanks holding 2 quarts (2 L) or more are adequate for limited-term breeding. Dur-

ing limited-term breeding the fish are not fed!

Long-term Breeding: The methods used here are basically the same as for limited-term breeding. The only difference is that the breeding fish stay together for a longer period (see descriptions, page 53).

This method is suited for species that lay eggs continually over a longer period but only a few at a time.

Breeding tanks for long-term breeding should hold at least 2.5 gallons (10 L). During long-term breeding the parent fish are fed several times a day (see Feeding of Breeding Stock, page 37).

Handling the Eggs

After the eggs have been deposited they must be stored until the baby fish hatch. How long they have to be stored before hatching and how warm they should be kept is indicated in the descriptions of the different species, starting on page 53. There are two ways of storing the eggs.

Dry Storage

The easiest and safest method of storing fish eggs is to keep them in damp peat. Because bacteria do not thrive in peat, the eggs are not as likely to get moldy. Also, because the eggs are spaced away from each other, any fungi developing on eggs that have died are unlikely to spread to healthy eggs. You therefore need not remove the spoiled eggs, which turn a milky white and are thus easy to recognize. Another advantage of this storage method is that all the fry hatch at the same time and thus develop at the same rate.

If you have used peat as a spawning aid, you can remove the peat fibers (in the case of plant spawners) from the breeding tank with your hands

Breeding Killifish

Storing fish eggs. The eggs of plant spawners can be kept in tight-shutting freezer containers. Place about ³/₈ inch (1 cm) of damp peat in the bottom, distribute the eggs individually on the peat, and top with a second layer of loose, damp peat.

or strain the mixed peat (in the case of bottom spawners) through a net. Squeeze the peat with your hands until no more water drips from it. Now check to see if there are any eggs. If a few of the eggs burst in the course of squeezing the peat or when you press them between your fingers, these either were not fertile or had not yet fully hardened. Let the peat mixture used by bottom spawners dry for one day, and then loosen it so that it is flaky and wrap it in three layers of newspaper. Store the package containing peat and fish eggs in a fish transport bag, and label the bag, writing down the name of the species, the starting date of storage, the number of eggs, and the date when the eggs are to be watered again (see Watering). Be sure the peat does not dry out during storage. Open the bag periodically to check the state of moisture. This also helps prevent the formation of mold.

If you have used a mop or a steel grate as a **spawning aid,** pick off all the eggs and store them. They are best kept in tight-shutting freezer containers (see drawing on left).

Important: Store the plastic bag or freezer container with the fish eggs inside in a Styrofoam box or an insulated bag. Enclose a thermometer so that you can check the storage temperature. Be sure to keep the eggs at the proper temperature for that particular species. If the eggs become too warm or too cool the embryos die or develop abnormally!

Watering the Stored Peat

After the fish eggs have incubated for the appropriate length, the peat in which the eggs are packed must be soaked again with the kind of water used in the breeding tank.

The right time for watering is easy to determine. When the surface of the eggs, which is shiny at first, begins to become cloudy and the eyes of the young fish (golden iris) are easy to spot, the time has come.

The temperature of the water used should always be 59–65° F (15–18° C). In this fairly cool water the fry hatch quickly and are healthy. Higher water temperatures usually result in fry that are not viable, the so-called belly crawlers.

Peat with eggs from plant spawners is soaked in small tanks or dishes in water 2 inches (5 cm) deep.

Peat with eggs from bottom spawners is mixed with water and stirred several times by hand in containers of adequate size (which depends on the amount of peat). Wait until the eggs and some of the peat have settled on the bottom. Skim off any peat floating on the surface. Then adjust the water level so that it is about 1¼–2 inches (3–5 cm) above the peat that has settled.

Tip: For killifish species whose eggs need a relatively long incubation time (see descriptions, page 53) it has proven beneficial to lower the storage temperature somewhat and lightly moisten the peat for a few days before watering. Once the water is added, the fry hatch rapidly all at once.

Breeding Killifish

Wet Storage

For wet storage the fish eggs are kept in water until the fry hatch. This method is used only for killifishes that are plant spawners.

Storage: Place the fish eggs in a dish (1 pint to 1 quart or 0.5–1 L) filled with fresh water that is the same as the water in the breeding tank. Place the dish, protected from sun and bright light, on or next to the breeding tank.

Unfortunately, large portions of eggs incubated this way frequently succumb to fungi. To prevent this, add medication to the water that inhibits fungus growth (available at pet stores). I have had good success with FUNGI-STOP (Tetra). In persistent cases, chlortetracycline is effective (see Directions for Using Drugs, page 43).

Check the dish daily, and remove fungus-infested and infertile eggs immediately. Scoop up fry that have hatched, and place them in a rearing tank with water of the same properties.

Note: If all the eggs succumb to fungus, it may be because the male parent was too young or infertile. Generally, substituting a different male for breeding takes care of the problem.

Aids for Hatching: If the fry have trouble hatching, you must help things along, for fully developed embryos die quickly if they cannot hatch. You can induce hatching as follows:

• Place the dish with the eggs in the refrigerator for 10–15 minutes.

• Then blow some air from your mouth into the water through a straw. This reduces the oxygen content of the water.

By doing this you imitate processes that trigger hatching in nature. In nature, the eggs are exposed to oxygen in the air during the dry season. When the rainy season comes, the bodies of water fill up again. Because of this and through decomposition processes (bacterial activity), the oxygen level drops dramatically, and this triggers the hatching of the fry. In addition, there is a general cooling at this time in nature. Often all that is required for the fry

of plant spawners to hatch is a thorough shaking of the eggs in the water.

Rearing the Fry

Once they have hatched, the fry are skimmed off with a spoon and transferred to a rearing tank.

The *rearing tank* that is used first should hold at least 2.5 gallons (10 L) and have an internal foam filter (see page 19). As the fry grow larger, they must be moved to larger tanks. For reasons of sanitation, no substrate is added to these tanks, but the presence of fast-growing aquatic plants (hornwort, and pond weed) is useful because the plants provide hiding places and keep the nitrate concentration low.

The *water* must be changed weekly (up to 70%) for species that grow slowly (see descriptions, page 53). For fast-growing fry, the water must be replenished at least every other day or growth may be stunted. Introduce the fresh water slowly because

Mating behavior. Shown here is a *Cynolebias nigripinnis* male wooing his chosen female with a sinuous dance.

Breeding Killifish

baby fish can be extremely sensitive to changes.

Feed the fry several times a day (see Food for Young Fry, page 37, and descriptions, page 53).

Leftover food and feces must be siphoned out of the water daily. *Ampullaria* snails are useful for cleaning up the rest.

Note: For rearing delicate fry, adding a teaspoonful of *Aureomycin* (available at pet supply stores) per 2.5 gallons (10 L) of water has proved effective against parasites. If annual fish are to be reared in considerable numbers, a teaspoonful of table salt added to each 2.5 gallons (10 L) of water does no harm and largely prevents velvet disease (see page 40), which is otherwise common in *Nothobranchius* species. *Aureomycin* can be added.

Tip: I consider it unnecessary to segregate fry by sex for rearing. If you keep them separately, the females of some species tend to grow excessively large but others grow abnormally slowly and never spawn. The sexes should be raised separately only if excessive biting makes this mandatory.

Deterioration Due to Inbreeding

The offspring of fish used over and over for breeding—without the introduction of new blood—often show genetic damage. This may take the form of

• physical malformations
• susceptibility to disease
• sterility
• lopsided sexual distribution of the offspring

Deterioration due to inbreeding may also be the reason for the sudden, unexplained disappearance of some aquarium species that used to be quite common. However, genetic deterioration brought on by inbreeding can be averted.

To Prevent Genetic Deterioration

• Keep at least three pairs—preferably more—for breeding.

• Cross in new specimens of the species regularly, if at all possible. It is crucial that crossbreeding of populations from different geographic regions be avoided.

• If you do not know the geographic origin of a species (this is often the case with *Aphyosemion australe* and *A. spoorenbergi* and with many strains that have been bred in aquariums for a long time), cross in only fish that are identical in appearance with those you already have.

Popular Killifish Species

In the preceding chapters you have learned the most important facts about killifish in general. The descriptions in this chapter will give you additional basic information about the distribution, life patterns, and reproductive behavior of individual species in nature, as well as their requirements in an aquarium and the necessary conditions for breeding them. By applying this information you can avoid gross mistakes in the maintenance and the breeding of killifish. The advice and tips given here are based on my own experience with this fish family. Of course, other methods can lead to success, too, but many discussions with successful keepers and breeders of killifish have confirmed my view that such differences in method are usually not very substantial.

The scientific names used in the descriptions are given in abbreviated form, and the measurements in parentheses after the names refer to the size of fully grown fish.

Aphyosemion: Bottom Spawners

This group of fishes, which have been combined under one heading on the basis of common reproductive behavior, includes small as well as very large killis (photograph on front cover and page 45). The majority of them are easy to keep and breed. The very large ones in particular, however, require a lot of attention as well as some luck to maintain and breed successfully.

Species: *A. filamentosum* (lyretail from Togo) (1.6–2 inches or 4–5 cm), *arnoldi* (Arnold's lyretail) (2 inches or 5 cm), *kunzi* (3.2 inches or 8 cm), *fallax* (3.6 inches or 9 cm), *sjoestedti* (golden pheasant Gularis;) (up to 5.5 inches or 14 cm), and others.

Distribution: Coastal areas in West Africa, from Togo to Cameroon. The subgenus *Raddaella* (*A. kunzi, splendidum,* and *batesii*) occurs on the inland plateau of Cameroon and Gabon.

Habitat: Primarily warm rainforest swamps in coastal areas. *Raddaella* in relatively cool rainforests, sometimes in tiny puddles of water.

Life Pattern: Bottom spawners; typical annual fish.

Food: Flying insects, water insects, worms; large species also eat small fish and shrimp.

Reproduction: The eggs are deposited in the bottom and survive the dry period there. Large numbers of eggs are produced. The fry grow very fast, and the young fish are often sexually mature at six weeks.

Maintenance

Maintenance tank: Depending on the size of the species, 2.5 gallons (10 L) or more; very large species need tanks of over 25 gallons (100 L). All species are suitable for community tanks with other fish of appropriate size. An exception is *Raddaella*. Tank decoration should offer hiding places among plants, especially for females. Species tanks should have about $3/4$ inch (2 cm) of peat at the bottom and include floating plants.

Water: 4–20° dH; pH 6–7.5; 75° F (24° C); for *Raddaella*, 64–72° F (18–22° C).

Feeding: Live, frozen, and dry food. *Raddaella* species are fastidious eaters. Fully grown fish should not receive rich food (no beef heart or mussel meat) or they became obese.

Breeding

Breeding tank: These fish can be bred in a species tank of 2.5 gallons (10 L) or more; use a spawning container with peat (see page 49) or add peat ($3/8–3/4$ inch, 1–2 cm, deep) to the bottom of the tank.

Water: 4–10° dH; pH 6–7; 75° F (24° C); for *Raddaella*, 64–72° F (18–22° C).

Feeding parent fish: Live food.

Breeding method: Long-term breeding in a species tank. Members of aggressive species (*A.*

sjoestedti and *fallax*) should be segregated by sex before the planned breeding and brought together for two days when the females show a full set of eggs. Always combine one male with two to three females if the species is aggressive.

Handling eggs: In long-term breeding, remove the peat every two weeks, dry it (see page 50), and store at 75–86° F (24–30° C). The eggs should be stored for six to eight weeks for *A. filamentosum*; up to six months, for *A. sjoestedti* and other large species; three to six months for *A. arnoldi*, and about three months for *Raddaella* species.

Rearing: Immediately start feeding with Artemia nauplii (watch out for cannibalism!). The fry grow very fast and must be fed several times a day. Give them plenty of room and change the water frequently.

Special remarks: Lower the water temperature in the breeding tank by about 5° F (3° C) before introducing the parent fish. This has an invigorating effect on the fish that helps them spawn better. It simulates rainfall in nature.

Bottom spawners. These killifish lay their eggs on or in the bottom material (substrate). Right after spawning, both the male and the female push off from the ground vigorously and in doing so hurl the eggs into the substrate.

Aphyosemion: **Plant Spawners**

The species mentioned here belong to several different groups, but they all have the same reproductive pattern in common (photographs on pages 63 and 64). A distinction is made between strains from the cooler rainforests of the high plateaus and those of the warmer savannas and coastal rainforests. Closely related species are grouped together.

Species: The species groups *A. cameronense, ogoense, striatum,* and *australe* (Cape Lopez lyretail, or lyretailed Panchax), *congicum,* (Blue Gularis) *bivittatum* (two-striped *Aphyosemion*), and many others. All of them (with few exceptions) grow to about 2 inches (5 cm).

Distribution: Central and West Africa, Ivory Coast to Zaire.

Habitat: Bodies that carry water all year.

Life Pattern: Solitary; stay in areas of still water beneath floating leaves and other aquatic plants.

Food: Flying insects, and water insects.

Reproduction: Plant spawners. The eggs are deposited on aquatic plants or on shore grass hanging into the water, or they develop in the water. They can survive a short dry period (up to three weeks).

Maintenance

Maintenance tank: Species tank for one pair or for a trio, at least 2.5 gallons (10 L) or more; a tank for several species, 7.5 gallons (30 L) or more. Also community tanks of 20 gallons (80 L) or more, together with characins, armored catfish, and dwarf cichlids. Use dark gravel or sand for the substrate, and provide some scattered plants that offer hiding places. Species from the rainforest like moderate light, and the water surface should therefore be covered with floating plants.

Water: 4-–5° dH; pH 6.5–7; 70–73° F (21–23° C) for fish from rainforest areas; 75–79° F (24–26°

Popular Killifish Species

C) for those from savannas.

Feeding: Live and frozen food; I advise against dry food.

Special Remarks: It is practically impossible to tell females of related species apart; each species should therefore be kept separate.

Breeding

Breeding tank: The aquarium should hold about 2.5 gallons (10 L), but breeding is also possible in bowls with about 2 quarts (2 L) of water. Use a mop or a handful of peat fibers as spawning substrate; for long-term breeding, supply several spawning mops, which also serve as hiding places for the females.

Water: 4–6° dH; pH 6–6.5; 68° F (20° C) for fish from rainforest regions; 75°F (24° C) for those from savannas.

Breeding method: Short-term; separate parent fish after 1 day. Long-term breeding with spawning mops is also possible; place a pair or a trio into a tank.

Feeding parent fish: Live or frozen food.

Handling eggs: Collect eggs and incubate in dishes. Keep them in water or damp peat fibers at 68° F (20° C) 75 or (24° C). Eggs kept in peat (for two to three weeks) are not as likely to develop fungus.

Rearing: Feed with Artemia nauplii and microworms. Rearing can be problematic. The fry (especially those of the *A. cameronense* group) may react badly to even slightly dirty water. Change the water frequently, and offer plenty of swimming space and hiding places. Do not start feeding Cyclops too early, or you may lose some of the fry.

Special remarks: For beginners, fish of the groups *A. striatum* and *A. bivittatum* are especially recommended. The exceptionally beautiful killis of the groups *A. cameronense* and *A. ogoense* are often problematic to raise.

Aphyosemion (Subgenus *Diapteron*)

These small killifish are easy to tell apart at a glance from all other *Aphyosemion* species (photograph on page 64). The colors, fins, and position of the fins all have a typical look. Unfortunately, *Diapteron* killis are often hard to breed and are consequently available only rarely.

Species: *A. (D.) georgiae, cyanostictum, abacinum, fulgens,* and *seegersi.* All grow to between 1.4 and 1.8 inches (3.5–4.5 cm).

Distribution: Primarily in the Ivindo Basin in Gabon.

Habitat: Shaded streams in the rainforest; cool water (around 68° F or 20° C) that is very acid and has no measurable hardness.

Life Pattern: The fish are shy and solitary.

Food: Primarily flying insects.

Reproduction: Typical plant spawners.

Maintenance

Maintenance tank: Species tanks only; at least 5 quarts (5 L)—better yet, 10 quarts (10 L)—for one pair. The aquarium should look dark. Use peat for the substrate, and plant thickly with Java fern, Java moss, or *Bolbitis.* No artificial lighting; weak daylight or indoor lights suffice.

Water: 4–10° dH; pH 6–6.9 ; 64–72° F (18–22° C). I "winter over" my *Diapteron* at 61–64° F (16–18° C).

Feeding: Live and frozen food, especially Cyclops, flies, and mosquito larvae. Never use Tubifex worms! In a pinch, use grindal worms, but it is better to let the fish go hungry for two days. Artemia nauplii are also eaten readily by mature *Diapteron*.

Special remarks: The males of all *Diapteron* species are extremely aggressive toward each other. I keep them separated, combining one or two females with a male.

Popular Killifish Species

Breeding

Breeding tank: 5–10 quarts (5–10 L); add a thin layer of peat to the bottom and let a mop hang down from the water surface.

Water: 4–10° dH; pH 6–6.9; 64–72° F (18–22° C).

Breeding method: Long-term.

Feeding parent fish: Live or frozen food, especially Cyclops and Artemia; also black mosquito larvae.

Handling eggs: Pick eggs out of the mop every day, and incubate them in dishes with fresh water (water quality the same as in the breeding tank). The fry hatch after 12 days. If eggs are not collected, some of them hatch, but picking the eggs off is more productive.

Rearing: Scoop out fry that have hatched in dishes of water immediately, and raise them segregated by size. Because of the aggressive nature of the males I combine pairs or trios right away, which are then not separated again. Fry that have hatched in breeding tanks within five days of each other can be raised together. Differences in size often lead to cannibalism. Freshly hatched fry start eating Artemia nauplii right away but grow very slowly. *Diapteron* are sexually mature at nine months and reach their full size at about one year.

Special remarks: Breeding these fish is only for experienced killifish fanciers! If the water temperature reaches 75–77° F (24–25° C) *Diapteron* stop reproducing and their metabolism slows down. If kept under optimal care, *Diapteron* live to over four years.

Aphyosemion (Subgenus *Paraphyosemion*)

The species included in this group of very colorful killifish live in waters that offer highly different conditions (ranging from those typical of rainforests to those found in dry savannas) (photographs on pages 27 and 28). This must be taken into account in the care of the fish.

Species: *A. gardneri* (steel-blue *Aphyosemion*), *a. mirabile, cinnamomeum, spoorenbergi, ndianum, puerzli,* and *amieti*. The males of all species measure $2^3/_4$ inches (7 cm) on the average; the females of all species are somewhat smaller.

Distribution: Waters in the coastal plains, the rainforests, and the savannas of Nigeria and western Cameroon.

Habitat: Streams, rivers, and swamps.

Life Pattern: Oriented more toward the bottom; semiannual.

Food: Any live food animals that are not too big.

Reproduction: Plant and bottom spawners. The eggs survive dry periods of up to six weeks. In water, the fry hatch about four weeks after having been spawned.

Maintenance

Maintenance tank: Species tank of least 5 gallons (20 L) or more; community tank of at least 20 gallons (80 L). River gravel or sand as substrate, some plants, and moderate to bright lighting.

Water: 4–15° dH, better on the low side; pH 6–7.5; for fish from rainforests (*A. gardneri lacustre, mirabile* [including subspecies], *cinnamomeum,* and *ndianum*), 68–75° F (20–24° C); for other species, 73–81° F (23–27° C).

Feeding: Live, frozen, and dry food. Feed generously during the growth phase, but give no more beef heart after the fish reach 2 inches (5 cm) or they tend to become too fat. Give lots of water fleas instead.

Special remarks: Excellent fish for beginners; well suited for community tanks. Males are aggressive toward each other, females less so. With good care *Paraphyosemion* can be expected to live for over three years.

Popular Killifish Species

Breeding

Breeding tank: Around 2.5 gallons (10 L) for short-term breeding; at least 5 gallons (20 L) for long-term breeding. Supply a spawning container with peat or peat fibers as spawning substrate (see page 47); also floating plants for females to hide in.

Water: 4–10° dH; pH 6–7.0; temperatures as given under Maintenance.

Feeding parent fish: Feed generously with live and frozen food.

Breeding method: Short or long-term. Separate parent fish about a week before the planned breeding. In short-term breeding, remove parent fish after two days.

Handling eggs: With long-term breeding, remove the peat once a week; with short-term breeding, take the peat out at the same time as the parent fish. Store eggs in peat. The fry hatch after 25–30 days at about 77° F (25° C).

Rearing: Start feeding with Artemia nauplii immediately. The fry grow fast and are quite hardy.

Special remarks: *A. ndianum* and *A. spoorenbergi* are harder to breed and raise.

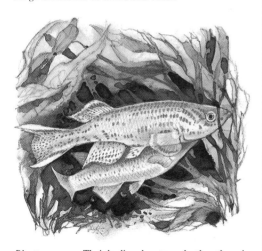

Plant spawners. Their bodies close to each other, the pair swims up to the plant on which the eggs will be deposited. A short time later the fish spawn.

Aplocheilus

Now that the rice fishes (*Oryzia*) have been assigned a family of their own, *Aplocheilus* species are the only killifish from the Indio-Malayan region (photographs on page 64). The larger species in particular have been some of the most popular aquarium fish for several decades.

Species: *A. lineatus* (*Panchax lineatus* or striped *Panchax*) (up to $4^3/_4$ inches, 12 cm); *A. panchax* (blue *Panchax* or *Panchax panchax*) ($2^3/_4$ inches, 7 cm); *A. blockii* (dwarf *Panchax,* green *Panchax,* or *Panchax* from Madras) (2 inches, 5 cm); *A. kirchmayeri* (2 inches, 5 cm), and *A. dayi* (Day's *Panchax* or Ceylon *Panchax*) ($2^3/_4$–$3^1/_2$ inches, 7–9 cm).

Distribution: From India to Thailand and Indonesia; *A. dayi* only on Ceylon.

Habitat: Rice fields and swamps; also streams and lakes.

Life Pattern: Singly in still waters among aquatic plants.

Food: Primarily flying insects (ants and others), but also water insects.

Reproduction: Spawn among plants close to the water surface; plant spawners.

Maintenance

Maintenance tank: For small species, tanks of 5 quarts (5 L) or more will do for one pair, but 5 gallons (20 L) or more are better; large species need tanks of over 13 gallons (50 L) (because of aggressiveness!). Any kind of substrate is suitable (surface fish); dense planting around the edges with fine-leafed plants (plant center and foreground sparingly); the surface should be partially covered with floating plants; moderate to bright lighting.

Water: 4–25° dH; pH 6–7.5; 75–86° F (24–30° C).

Feeding: Live, frozen, and dry food; preferably flies and mosquito larvae. *A. lineatus* also take crickets and earthworms.

Special remarks: Excellent fish for beginning hobbyists; well suited for community tanks. These fish love to jump; the aquarium must therefore be tightly covered.

Breeding

Breeding tank: About 2.5 gallons (10 L); for *A. lineatus,* 5 gallons (20 L) or more. Provide a spawning mop that hangs down from the water surface or fine-leafed aquatic plants for spawning on.

Water: 4–15° dH; pH 6–7; 81° F (27° C).

Feeding parent fish: Live and frozen food.

Breeding method: Separate parent fish for about one week before breeding; short-term breeding for one day works best. Place the fish in the breeding tank in the evening, and remove them the next day around noon. Long-term breeding is possible, too.

Handling eggs: Either leave the eggs in the tank or collect them and place them in bowls with the same quality water. In long-term breeding, pick out the eggs every day. The fry hatch after 12–14 days in water of 81° F (27° C).

Rearing: Start feeding with Artemia nauplii and microworms at once; rearing is unproblematic.

Special remarks: Be very careful in selecting breeding stock; color often fades and size diminishes in aquarium-bred offspring.

Epiplatys and *Pseudepiplatys*

African *Epiplatys* have been popular aquarium fish for some time. They are reminiscent in shape of pikes. The large, undershut mouth indicates that the species in this group are surface-dwelling fish.

Species: *E. dageti* (2.4 inches or 6 cm), *E. sexfasciatus sexfasciatus* (six-banded *Panchax*) (3.2 inches, 8 cm), and *E. sexfasciatus rathkei* (3.6 inches, 9 cm). At present about 40 species are assigned to this genus. *Pseudepiplatys annulatus* (1.6 inches, 4 cm) is the only species in that genus.

Distribution: Found over a wide range from Gambia to the headwaters of the Nile and as far south as the Congo.

Habitat: Streams in rainforests, savannas, and coastal plains.

Life Pattern: Lying in wait for prey at the water surface; not very sociable, but not solitary.

Food: Flying insects, which are caught on the surface of the water and by jumping into the air.

Reproduction: Plant spawners; normally the eggs are attached to aquatic plants, but some species also spawn on the bottom.

Maintenance

Maintenance tank: Species tank with 5 gallons (20 L) or more for one pair; community tanks at least 20 gallons (80 L). There should be plenty of room for swimming. Use river gravel, sand, or crushed basalt or lava for the substrate; plant densely along edges (sparsely in the center and foreground), and add floating plants; moderate to bright lighting.

Water: 4–15° dH; pH 6–7; 72–79° F (22–26° C).

Feeding: Primarily live and frozen food, such as fruit flies and mosquito larvae, or small crickets (only for large species); also dry food.

Special Remarks: Many species are suitable for beginners.

Breeding

Breeding tank: 2.5 gallons (10 L) or more; supply a spawning mop or a handful of peat fibers; for long-term spawning, supply several mops that also offer hiding places for the females.

Water: 4–10° dH; pH 6–7; 72–75° F (22–24° C).

Feeding parent fish: Only live and frozen food, such as mosquito larvae, water fleas, and flies.

Popular Killifish Species

Breeding method: Short-term breeding is more productive, but long-term breeding is also possible.

Handling eggs: In long-term breeding, pick off the eggs every day and store them wet or dry at 72–75° F (22–24° C) for 12–14 days.

Rearing: Start feeding microworms and Artemia nauplii immediately.

Special remarks: Raising *Pseudepiplatys annulatus* is more difficult. The water must be soft (4–5° dH) and slightly acid (pH 6–6.5). The real problem is the fry, which are so tiny that they can hardly manage to eat slipper animalcules (*Paramecium*). Wheel animalcules (Rotifera) are best used as rearing food. Long-term breeding in a very thickly planted tank works best. There the fry find enough microscopic food for the first few days. Once the fry can be seen swimming among the plants, Artemia nauplii can be given, which are also consumed by the adult fish. Scoop out the baby fish from time to time, and rear them in a thickly planted tank.

Killifish of North America and Europe with Special Requirements

Not all the killis discussed here come from environments with extreme conditions, but the great majority are adapted to unusual habitats. Thus, many species live in brackish water or in waters along seacoasts; others even dwell in salt lakes or mineral springs. Desert fish (*Cyprinodon*) are famous for surviving extreme temperature fluctuations and water qualities.

Species: *Lucania goodei* (2 inches, 5 cm), from Florida; *Jordanella floridae* (American flagfish) (2.4 inches, 6 cm); *Profundulus punctatus* (3.6 inches, 9 cm) from Mexico and Guatemala; *Cyprinodon variegatus* (1.6–2.4 inches, 4–6 cm), from the Caribbean; *C. macularius* (2.4 inches, 6 cm) from California and Mexico; *Fundulus heteroclitus* (4.7 inches, 12 cm) from the East Coast of the United States; *Aphanius iberus* (Spanish minnow) (2 inches, 5 cm); *A. mento* (Persian minnow) (2 inches, 5 cm) from Turkey to Syria.

Distribution: *Aphanius* in the Mediterranean, also in oases. *Cyprinodon* primarily in desert areas in the United States and Mexico. *Fundulus* and other genera, some of which are extremely rare, from Newfoundland to southern Central America.

Habitat: *Jordanella floridae, Lucania goodei,* and *Profundulus punctatus* live in freshwater; the other species inhabit slightly salty to very salty waters mostly with high temperatures. The northernmost representatives are cold-water fish.

Life Pattern: Extremely varied; the fry of desert fish sometimes live in shoals, but otherwise these fish are solitary.

Food: Omnivorous; many species also eat algae.

Reproduction: Plant spawners; eggs survive dry periods only in rare cases (*Fundulus confluentes*). Reproduction is often restricted to certain periods or spawning cycles.

Maintenance

Maintenance tank: Species up to 2 inches (5 cm) in tanks of at least 2.5 gallons (10 L) with one pair per tank. For saltwater species, use tanks of at least 8 gallons (30 L). Large species, such as *Fundulus heteroclitus,* require tanks of 25 gallons (100 L) or more. Sandy bottom or coarse gravel; plant with pond weed and hornwort. Filter well, with additional aeration if necessary. Normal to bright lighting (desert fish). Treat sweet-water species like *Aphyosemion* (see page 54).

Water: 8–15° dH; pH 7.5–8.5; 72–79° F (22–26° C); for desert fish, 79–91° F (26–33° C). Many species from northern areas and from the Atlantic coast need a "cold season"; these fish should, if possible, be wintered over in the cellar.

Popular Killifish Species

The tank water should have 3 tablespoons of sea salt added per 2.5 gallons (10 L); seawater or water with higher salinity is also suitable to keep these species in.

Feeding: Live, frozen, and dry food, plus vegetarian food like algae, spinach, and lettuce.

Special remarks: Saltwater species, especially some members of the *Aphanius* genus, can also be kept in a marine aquarium.

Breeding

Breeding tank: 5 gallons (20 L) or more; for large species, 13 gallons (50 L) or more. Sandy bottom or a stainless steel grate. Sweet-water species can spawn on mops or Java moss.

Water: See Maintenance; changing the water stimulates spawning.

Feeding parent fish: Live and frozen food, with additions of vegetarian food.

Breeding method: Segregate parent fish before spawning. For sweet-water species, follow directions for breeding plant spawners (see page 47). Saltwater species all prey on their eggs. Allow them to spawn in fine sand, and then sift out the eggs with a net. A steel grate with some green perlon wadding under it can also prevent the parent fish from eating the eggs. Always set up breeding fish in single pairs!

Handling eggs: Allow them to incubate in bowls of water with the same properties as the water in the breeding tank. For temperature, see Maintenance. Length of incubation for *Cyprinodon*: six to ten days; other species take 14 days or more.

Rearing: Spacious tanks; use *Artemia* nauplii as the first food.

Special remarks: Species from temperate latitudes with strong seasonal fluctuations spawn only at certain times of the year (spring). These fish (*Fundulus heteroclitus*, for instance) must be wintered over in cool temperatures.

Lamp-eyes

The members of this subfamily are easily told apart from all other killifish species by their bright, iridescent eyes. In addition, most lamp-eyes are shoal fish (photographs on page 60).

Species: Of the various species the following are commonly offered for sale: *Plataplochilus ngaensis* and *P. miltotaenia* from Gabon, both measuring about 2 inches (5 cm); *Aplocheilichthys macrophthalmus* (1.4 inches, 3.5 cm), from Nigeria, *Lamprichthys tanganicus* (up to 6 inches, 15 cm) and *Procatopus aberrans* (2.2 inches, 5.5 cm), from Cameroon and Nigeria. The females of all these species are up to $^1/_3$ smaller than the males.

Distribution: West Africa to southern Central and East Africa. *Lamprichthys* live only in Lake Tanganyika.

Habitats: Very varied, reflecting the great number of species and the large area of distribution. Primarily, however, streams in rainforests and in savannas.

Life Pattern: Typical shoal fish; the most colorful species are solitary.

Food: Water insects and flying insects.

Reproduction: Spawn in pairs among plants near the surface of the water; also in cracks formed by roots and rocks.

Maintenance

Maintenance tanks: Shoals of up to 10 fish in species tanks of 13 gallons (30 L) or more. *Lamprichthys* should have tanks of no less than 25 gallons (100 L) and preferably 50 gallons (200 L) or more; well suited for community tanks. Use sand as substrate; plant edges densely (sparsely in the center and foreground), add a few floating plants, and supply normal lighting. Filtering with not too much current and some still zones. Plenty of root wood and, for *Lamprichthys*, some rock structures.

Water: 4–15° dH; pH 6–7; 68–75° F (20–24°

Popular Killifish Species

C) for *Plataplochilus*, up to 79° F (26° C) for *Aplocheilichthys macrophthalmus*. *Lamprichthys* species must have water with 15–25° dH hardness, a pH of 7.5–8.5, and a temperature of 73–77° F (23–25° C). Make sure the water is always clean!

Feeding: Live and frozen food, especially flies, ants, water fleas, but no Tubifex worms; also dry food.

Special remarks: Lamp-eyes, especially those of the *Lamprichthys* genus, tend to develop fungus infections if the skin is injured. To remove these fish, always use soft, fine-meshed nets or a glass trap.

Breeding

Breeding tank: See Maintenance; make sure the water is clean. No plants; only a spawning mop, a pumice stone with grooves filed into it, or a floating cork square.

Water: Fresh and especially clean. For *Lamprichthys*, see under Maintenance; for other kinds: 4–10° dH; pH 6–7; for temperatures, see Maintenance.

Feeding parent fish: Only live and frozen food! Feed sparingly so that the water quality does not deteriorate too much.

Breeding method: Long-term spawning over a period of several days. Place more females in the tank than males.

Handling eggs: Pick the eggs off the mop daily and incubate in dishes (2 quarts, 2 L). Add medication to prevent fungus. Change pumice stones or corks every day and place them, too, in dishes, floating.

Rearing: Scoop hatched fry out of dishes every day, and rear them, segregated by size, in large tanks. Replace part of the water frequently, and siphon off leftover food daily. First food: vinegar eelworms, microworms, and freshly hatched *Artemia* nauplii. Very small species should be fed slipper animalcules for the first few days.

Special remarks: Make sure the water is always fresh and clean. If there is any sign of illness in the fish (fins clamped to body), change the water at once.

Nothobranchius

The killis of this group are among the most brilliantly colored of all aquarium fish (photograph on page 46). Unfortunately their vitality is largely exhausted after six months, but their behavior and the beauty of their colors make up for this drawback.

Species: *N. rachovii* (Rachow's nothobranch) (2.4 inches, 6 cm), *N. melanospilus* (Beira nothobranch); (2.8 inches, 7 cm), *N. guentheri* (Guenther's nothobranch), *N. foerschi*, and others.

Distribution: Savannas of Central and East Africa.

Habitat: Puddles formed by temporary flooding and offering conditions that fluctuate widely with the seasons.

Life Pattern: The fish swim around actively in a constant search for food.

Food: Anything alive that the fish can manage to swallow.

Reproduction: Typical annual fish; the eggs are deposited in the bottom and must go through a dry period in order to develop.

Maintenance

Maintenance tank: Species tanks of 8 gallons (30 L) or more (for five pairs); community tanks of 20 gallons (80 L) or more, in which these fish are combined with others of comparable size and peaceful disposition, such as lamp-eyes, *Paraphyosemion*, and African barbs. River gravel or sand is used as substrate, and the rear is densely planted. Provide shady area through floating plants. Normal to bright lighting. Decorate with root wood.

Water: 4–25° dH; pH 6–7; 79° F (26°C).

Popular Killifish Species

Feeding: Live and frozen food, especially Tubifex worms, red mosquito larvae, grindal worms, and beef heart; also dry food.

Breeding

Breeding tank: At least 5 quarts (5 L) for a trio. Put a layer of $^3/_8$– $^3/_4$ inch (1–2 cm) of peat in the bottom of the tank.

Water: 4–15° dH; pH 6–7; 79–86° F (26–30° C).

Breeding method: Long-term.

Feeding parent fish: Live and frozen food, such as mosquito larvae and Tubifex worms, plus water fleas for roughage.

Handling eggs: Remove peat from tank once a week, allow to dry for a day wrapped in newspapers, and then store at 77–86° F (25–30° C). The length of incubation depends very much on the moisture content and temperature of the peat. For red-tailed species, it takes about three months; for *N. rachovii*, 4–6 months.

Rearing: Artemia nauplii; if the young fish are fed several times a day, they can reach sexual maturity as early as at four weeks.

Special remarks: Red-tailed species (*N. guentheri, N. foerschi,* and others) can be bred without difficulties. The fry of *N. luekei* and *N. janpapi* must be fed strained vinegar eelworms at first (see page 39). *N. janpapi* spawn in open water.

Rivulus: Large Species

Large *Rivulus* species are among the most rewarding and least demanding aquarium fish (photograph on page 18). They are not only excellent jumpers, but in their natural habitat also wander across considerable distances of dry land. In the wild they are often caught in tire tracks and puddles that are not near any bodies of water.

Species: The kinds most commonly kept in aquariums are *R. cylindraceus* (2.4 inches, 6 cm), *R. magdalenae* (2.8 inches, 7 cm), *R. amphoreus* (3.2 inches, 8 cm), *R. holmiae* (4 inches, 10 cm), as well as a species not yet described in scientific literature, namely *R. spec. aff. holmiae*, which can grow as large as 6.3 inches (16 cm). *R. caudomarginatus* and *R. ocellatus* live in the brackish water of lagoons.

Distribution: From Florida and Cuba across Central America and some Caribbean islands and in all of South America east of the Andes.

Habitat: Streams, ditches, swamps, and small bodies of water of all kinds.

Life Pattern: Singly beneath floating leaves of aquatic plants or below vegetation hanging over the water from the shore. Also seen sunbathing on floating leaves of aquatic plants.

Food: Flying insects.

Reproduction: Depending on the species, the eggs are deposited on the bottom, on plants, or on grass hanging over the water. *R. ocellanus* is self-fertilizing; that is, a single specimen is all that is needed for reproduction.

Maintenance

Maintenance tank: Community tanks of 20 gallons (80 L) or more for several pairs. Sand or gravel as substrate; thick planting with sword plants, *Ceratophyllum*, and water trumpets; moderate to bright lighting; root wood and aerial roots for decoration (see page 21).

Killifish of the genus Aphyosemion. *Above left: Two male* Aphyosemion mimbon *Medouneu. Above right: A male* Aphyosemion cameronense *Ngoyang. Middle left: A male* Aphyosemion ogoense pyrophore *RPC 18. Middle right: A male* Aphyosemion bivittatum *Funge. Below left: A male* Aphyosemion congicum. *Below right: A male* Aphyosemion (Diapteron) abacinum.

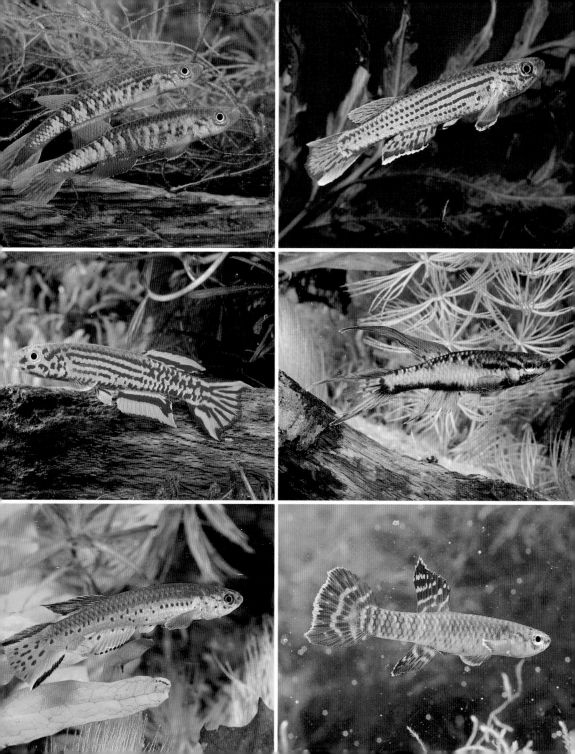

Popular Killifish Species

Water: 4–25° dH; pH 6–7.5; 72–79° F (22–26° C). For species that live in brackish water use hard water with added sea salt (3 teaspoons per 10 quarts); 75–86° F (24–30° C).

Feeding: Live, frozen, and dry food. Growth can be furthered by giving beef heart and mussel meat. *R. caudomarginatus* and *R. ocellatus* should be constantly "surrounded" by food. Allow water fleas and *Cyclops* to swim in the water, and give the fish *Tubifex* worms and dry food now and then.

Special remarks: Males fight among each other but never get seriously hurt if there is enough room to get out of the way. It is essential that the tank be tightly covered!

Breeding

Breeding tank: 2.5 gals. (10 L) or more, with peat fibers or a spawning mop hanging from the water surface.

Water: 4–15° dH; pH 6–7; 77–81° F (25–27° C).

Feeding parent fish: Live and frozen food.

Breeding method: Short-term; keep sexes segregated for two weeks, then bring them together for 24 hours. If the water temperature drops slightly, the fish spawn. Remove the fish after they have spawned.

Handling eggs: Collect eggs and incubate them for three to four weeks at 68–77° F (20–25° C) in damp peat.

Rearing: Start feeding immediately with *Artemia* nauplii and *Cyclops* several times a day.

Different killifish species. Above: A pair of Aplocheilus lineatus (Panchax Lineatus or striped Panchax); Below: A male Lamprichthys tanganicus (lamp-eye).

Rivulus: Small Species

The small *Rivulus* species are some of the most colorful killifish from South America (photograph page 18). These killis have become better known in the last few years because new kinds have been imported and aquarists have become more interested in them. *R. agilae* and *R. strigatus* are among the more popular varieties that are now stocked regularly by dealers.

Species: *R. agilae, R. geayi,* and *R. strigatus;* these species grow to 2 inches (5 cm). *R. punctatus* (1.6 inches, 4 cm), *R. ornatus* (1.4 inches, 3.5 cm), *R. xiphidius* (1.6 inches, 4 cm), and *R. luelingi* (2 inches, 5 cm).

Distribution: From Venezuela to Peru, most common in Guyana, Surinam, and French Guiana.

Habitat: Small rivers, streams, and flooded meadows.

Life Pattern: Solitary; males claim small territories.

Food: Flying insects, water insects, and worms.

Reproduction: The females pass through the territories of males and spawn on plants and roots.

Maintenance

Maintenance tank: 2.5 gallons (10 L) or more for one pair; several pairs need 13 gallons (50 L) or more (territories!). Use sand as substrate and plant thickly around edges (few plants in the center and foreground) to provide hiding places that are necessary for females. Provide good filtration with moderate water flow and some still areas. Normal daylight hours.

Water: 4–15° dH; pH 6–7; 72–77° F (22–25° C); very clean.

Feeding: Live and frozen food; also Tubifex worms; small amounts of dry food.

Special remarks: *R. luelingi* from southern Brazil needs lots of oxygen and low temperatures (64–70° F or 18–21° C).

Popular Killifish Species

Breeding

Breeding tank: 5 quarts (5 L) or more, with a spawning mop. The mop should be large enough that there is hardly any room left for swimming.

Water: 4–10° dH; pH 6–6.8; 75–79° F (24–26° C); *R. luelingi:* 64–68° F (18–20° C).

Feeding parent fish: Only live and frozen food; flies and moderate amounts of Tubifex encourage the production of spawn.

Breeding method: Short-term breeding gives the best results. Keep sexes segregated; then combine a female with a good set of egg cells with a male in the breeding tank in the morning. Remove the fish from the tank in the afternoon (otherwise they eat the eggs!).

Handling eggs: Pick out the eggs, and incubate them damp in peat or wet in dishes. The fry are ready to hatch after not quite three weeks. For incubation, see water temperature in breeding tank.

Rearing: Start feeding with Artemia nauplii immediately. Fry are sometimes sensitive to dirty water. Once the males begin to develop color, hiding places (plants) must be provided.

Special remarks: *R. agilae* and *R. strigatus* can be recommended for beginners. Other species, especially *R. xiphidius,* present problems.

Roloffia: Large, Bottom-spawning Species

The International Commission of Zoological Nomenclature has decided to include the genus *Roloffia* in the genus *Aphyosemion* (photograph on page 17). However, not all scientists agree with this decision, and I follow Etzel and van den Audenaerde [*DKG Journal (Journal of the German Killifish Society)* 20(7), November 1988] in continuing to use the name *Roloffia.*

Species: *R. occidentalis, R. toddi, R. monroviae,* and *R. huwaldi.* All species grow to between 3.2 and 4 inches (8–10 cm).

Distribution: West Africa, Sierra Leone, and Liberia.

Habitat: Shallow, standing water, as in rice fields and swamps.

Life Pattern: Aggressive fish that come together only to mate.

Food: Flying insects and *Culex* mosquito larvae.

Reproduction: Bottom spawners; typical annual fish.

Maintenance

Maintenance tank: Species or community tank of 15 gallons (60 L) or more for two pairs or trios. River gravel or sand as substrate; provide plenty of hiding places. Moderate lighting.

Water: 4–20° dH; pH 6–7.5; 72–79° F (22–26° C).

Feeding: Live and dry food; during the growing phase also beef heart and mussel meat, as well as dry food. For adult fish of large *Roloffia* species water fleas and freshwater amphipods are the best diet because they prevent obesity.

Special remarks: Fish of the large *Roloffia* species do not live long; the greatest age they have reached in my aquariums—if kept at 73–77° F (23–25° C)—is 6 months. But the young fish develop beautiful coloration as early as at 6 weeks if kept under optimal conditions, however. All *Roloffia* species are aggressive toward their own kind but do not bother other kinds of fish in the aquarium.

Breeding

Breeding tank: 5 gallons (40 L) or more for one trio; place $^3/_8$–$^3/_4$ inch (1–2 cm) of peat in the bottom of the tank, and add some floating plants.

Water: 4–10° dH, preferably on the soft side; pH 6–6.9; 75–79° F (24–26°C). When you change the water, the temperature may temporarily drop to 68° F (20° C).

Feeding parent fish: Live and frozen food,

especially mosquito larvae, freshwater amphipods, and flies.

Breeding method: Long-term.

Handling eggs: Remove the peat after the fish have been breeding for two to four weeks and allow it to dry wrapped in newspapers for a day, and then store it at about 77° F (25° C). The eggs will probably hatch after four to six months, although they may take nine months or more.

Rearing: Start feeding immediately with Artemia nauplii. Young fish can be combined for spawning as early as after six weeks.

Special remarks: Breeding these fish presents some problems. The best size for breeding is 2.4–2.8 inches (6–7 cm). Fish that have reached their full size are almost useless for breeding.

Roloffia: Small Species

The small *Roloffia* species are divided into two groups: the *R. petersii* group and the *R. liberiensis* group (photograph on page 17). The species associated with *R. liberiensis* are brightly colored. The differences between fishes within a group are gradual and therefore often hard to detect for nonspecialists.

Species: *R. petersii* and related species *R. guineensis, R. viridis, R. maeseni* ,and others. *R. liberiensis, R. roloffi*, and *R. geryi*, as well as some other, rare kinds. Size around 2 inches (5 cm); only a few species occasionally grow to more than 2.4 inches (6 cm).

Distribution: West Africa, from Gambia to Ghana.

Habitat: Small bodies of water from the coastal plain to the inland plateau.

Life Pattern: Shy fish that live singly in shallow zones near the edge of the water.

Food: Flying insects, but also worms.

Reproduction: Bottom-oriented plant spawners.

Maintenance

Maintenance tank: Species tank of 2.5 gallons (10 L) or more for one pair or trio; these fish are extremely shy in a community tank. Supply ³/₈–³/₄ inch (1–2 cm) of peat as substrate; aquatic plants that drift in the water are suitable for planting. Moderate lighting.

Water: 4–10° dH; pH 6–6.9; 70–73° F (21–23° C).

Feeding: Live and frozen food, preferably worms, water fleas, *Tubifex* (but not too much!), red mosquito larvae, and flies.

Special remarks: The different species as well as fish from different geographic origins must be kept strictly separate.

Breeding

Breeding tank: At least 5 quarts (5 L) per pair, preferably larger. Supply a handful of peat fibers or a mop as spawning substrate.

Water: 4–10° dH; pH 6–6.9; 73–75° F (23–24° C).

Feeding parent fish: Plenty of live and frozen food.

Breeding method: Short-term; keep the parent fish segregated for a week beforehand. These fish prefer to spawn near the bottom.

Handling eggs: For wet incubation, collect the eggs and keep them in dishes. For dry incubation, remove the peat and the eggs. The fry hatch after 14 days if kept at 72–75° F (22–24° C).

Rearing: Start feeding immediately with Artemia nauplii. The fry grow very slowly.

Special remarks: Select young fish about seven to nine months old for breeding. Fully grown fish are not suitable for breeding.

South American Bottom Spawners

In this group of killifish there are many varieties of shape and huge differences in size (from 1.2 to 7 inches, 3–18 cm) as well as in color (from "mouse gray" to the most brilliant shades!). (See photographs on inside front cover, page 28, and inside back cover). Many species are easy to keep and breed; others are decidedly difficult, such as *Rachovia pyropunctata* and *Campellolebias* species.

Species: *Cynolebias nigripinnis* (black-finned pearl fish) (2.4 inches, 6 cm), *C. citrinipinnis* (1.2 inches, 3 cm), *C. elongatus* (over 6 inches, 15 cm), and about 30 other species. *Pterolebias longipinnis* (longfin) (over 4.7 inches, 12 cm), *P. zonatus* (4 inches, 10 cm), and the genera *Trigonectes, Neofundulus, Campellolebias,* and *Austrofundulus* are only just being discovered by aquarists. *Pterolebias* spec. NSC 1 should be in a separate genus.

Distribution: South America, from Venezuela to Argentina.

Habitats: Bodies of water that dry up temporarily and are located in rainforests or savannas (Argentinian pampas and the Venezuelan llanos), in coastal plains, and on inland plateaus.

Life Pattern: The males of almost all the species defend their territory against intruders. The weaker males are suppressed or even killed if they cannot escape. Females have inconspicuous coloring and are hardly aggressive at all.

Food: Live food of all kinds; *C. elongatus* preys on other fish.

Reproduction: Typical annual fish; bottom spawners. A dry period is mandatory for the eggs to develop.

Maintenance

Maintenance tank: Species tanks of anywhere from 5 quarts to 25 gallons (5–100 L). Only species at least 2 inches (5 cm) long can be combined in a community tank with other, nonaggressive fish (characins and catfish). Very large species are better kept in a species tank. Use river gravel or sand as substrate; plant densely around edges (few plants in

Substrate divers. Many South American bottom spawners dive into the bottom mud the whole length of their bodies to spawn.

the center and foreground), and supply floating plants. Moderate lighting.

Water: 4–15° dH; pH 6–7.5; for fish from open waters in coastal plains and savannas and for those from Venezuela and the Amazon Basin, 77–81° F (25–27° C). Southern types from Argentina and Uruguay, 59–68° F (15–20° C). *C. nigripinnis* and *elongatus* manage with 50° F (10° C) or even less during the growing phase. Fully grown fish should be kept warmer.

Feeding: Live and frozen food; also dry food. Plenty of worms (grindal, Tubifex, and earthworms). *C. elongatus* eats several times its own weight every day.

Breeding

Breeding tank: Small species, 2.5 gallons (10 L) or more; large ones, from 7.5 gallons (30 L) up; very large ones (*C. elongatus,* spec. NSC 1), than 13 gallons (50 L). Use a spawning container with peat at the bottom (see page 47) and include floating plants in the tank.

Water: 5–10° dH; pH 6–6.5; for *C. elongatus, nigripinnis,* and similar fish, 72–77° F (22–25° C); fish from Venezuela and other warm regions, 81–86° F (27–30° C).

Feeding parent fish: Live and frozen food.

Breeding method: Long-term.

Handling eggs: Remove the eggs along with the peat; squeeze the peat well, and allow it to dry wrapped in newspapers for a day. Loosen the peat before packing the eggs. Incubate two to eight months at about 77° F (25° C). Wet the eggs of all species with water or 59° F (15° C). Warmer water often results in fry that is not viable.

Rearing: It is best to begin feeding with microworms and *Artemia* nauplii; the fry must be fed several times a day. The young fish grow very fast. Change the water daily!

Addresses and Literature

Societies

American Killifish Association
c/o Mr. Ronald Coleman
903 Merrifield Place
Mishawaka, IN 46544

British Killifish Association
c/0 Mr. A. Burge
14 Hubbard Close
Wymondham
Norfolk NR18 ODH
England

Magazines

Aquarium Fish
P.O. Box 6050
Mission Viejo, CA 92690

Freshwater and Marine Aquarium
144 West Sierra Boulevard
Sierra Madre, CA 91024

Tropical Fish Hobbyist
One T.F. H. Plaza
Neptune City, NJ 07753

Books

Killifish Master Index, AKA Publications, Mishawaka.

Terceira, T., *Killifish, Their Care and Breeding,* AKA Publications, Mishawaka, 1975.

Index

Index

Index